The Ideal Result

Jack Hipple

The Ideal Result

What It Is and How to Achieve It

 Springer

Jack Hipple
Innovation-TRIZ
Tampa, FL, USA

Additional material to this book can be downloaded from
http://extras.springer.com/2012/978-1-4614-3706-2

ISBN 978-1-4614-3706-2   ISBN 978-1-4614-3707-9 (eBook)
DOI 10.1007/978-1-4614-3707-9
Springer New York Heidelberg Dordrecht London

Library of Congress Control Number: 2012938707

Printed on acid-free paper

Springer is part of Springer Science+Business Media (www.springer.com)

Everyone who writes a book dedicates it to someone and this will be no exception. My wife, Cindy, has stood by me on my creativity and innovation journey for 44 years, including 16 geographic moves, the raising of four beautiful daughters, and dealing with company and position changes that, on many occasions, were challenging—to say the least. She has tolerated the ambiguity of learning new things, the wide variety of challenging career circumstances, and all the while allowing me to chase my dream of impacting the productivity and creativity of others. There is no other author who has had the degree of support and encouragement that I have had. To her this book is dedicated.

Introduction

As someone trained and experienced with many different innovation and creativity tools, processes, and techniques over the first 25 years of my industrial career, yet frustrated at their usefulness in many technical and engineering applications, I had a significant learning experience while working as a project manager for the National Center for Manufacturing Sciences in Ann Arbor, Michigan, in 1994.This was shortly after leaving Dow Chemical as its Director of Discovery Research and Director of Corporate Chemical Engineering R&D. Despite the exposure to psychologically based creativity tools such as brainstorming, Creative Problem Solving, Six Thinking Hats®, and Lateral Thinking®, I had not had any serious training in quality tools, so I decided to attend a general quality conference sponsored by the American Society for Quality (ASQ) in Detroit. I had no mission other than to learn some of the fundamentals of QFD and other tools related to quality improvement. Though I am not sure that I can remember what I had for lunch yesterday, I remember very clearly at 3 PM that Friday afternoon 18 years ago, looking at my watch, and considering going home early to beat the traffic on I-94 between Detroit and Ann Arbor. I looked at the program and saw listed a presentation by two Russian scientists entitled, "The Use of TRIZ to Resolve the Contradictions of Pizza Box Design". Now I don't know about you, but that title was simply too interesting to walk away from. Having eaten a lot of pizza in my lifetime, I had some familiarity with pizza boxes, but had no idea what cosmic issues might be involved in pizza box design, why it was being discussed at a quality conference, nor what "TRIZ" was. All of us have life changing events in our lives and this was one of mine. I often wonder today what I would be doing had I not stayed to listen to that talk.

This presentation, by two of the original Russian emigrants who brought the TRIZ problem solving process into the United States (and I am humbly apologetic for not having written their names down), was about an inherent contradiction in takeout pizza products. The customer wants the pizza to be hot, but hot pizza gives off steam. As a result, the cardboard box lid absorbs the steam, softens, and collapses down on the pizza. When the customer lifts the lid of the pizza box after arriving home, some of the pizza, which has stuck to the lid, attaches to the raised lid and the customer is not pleased because a significant amount of the cheese that he paid for is not edible. We want the pizza to be hot for one reason (an enjoyable pizza) and we want it to be cold for another (preventing the lid from becoming soggy and collapsing due to steam and water absorption). The little plastic tripod that is in the center of many takeout pizzas was the subject of the talk I heard. Such a simple invention! It *resolves* the contradiction of the pizza being hot and cold at the same time. The pizza box becomes more *ideal* (from the standpoint of the consumer) through the resolution of a contradiction. You have to ask yourself why it took years for someone to invent this simple, extremely inexpensive device. But then again, my experience (and possibly yours) is that breakthrough inventions very often look so simple in hindsight. In my TRIZ problem solving work since, I have yet to see a breakthrough solution that did not seem so simple, in hindsight, that everyone asked themselves, "Why didn't we think of that before?" We will see other such examples as we explore the world of TRIZ.

After hearing that talk, I resolved to learn more about the mental process that provided that solution through the resolution of the contradiction between hot and cold pizza. That journey introduced me to TRIZ (Russian acronym for the Russian phrase, "algorithm for solving inventive problems") meaning the solving of problems whose solutions required special inventiveness, whose solutions were not immediately obvious, or ones that contain a very difficult design or operability contradiction. Our normal approach to problems with inherent contradictions that at first glance seem intractable is to compromise around the problem's parameters, yielding a solution that minimizes the pain. These types of solutions improve the situation, but are far from truly optimal. I like to call these "temporarily solved" problems, because someone improves it slightly (puts a bandage on it rather than stop the bleeding), the problem is then passed on to someone else who also improves it slightly, and so on. Within a corporation, you hope you have a new job when it comes back around and you see your old problem now wrapped in "improvement" bandages. Consider the current design of TV remote control devices as an example.

This book is about my journey of understanding and explanation to others of Inventive Problem Solving. It embodies my approach to training and understanding of TRIZ. There are many other books on the topic written by authors for whom I have great respect, with whom I have collaborated, and who have taught me a great deal. All of them approach the topic of innovation and TRIZ in a slightly different manner. Many are highly "left brained" and focus on detailed TRIZ algorithms, which are very useful and can be used for complex problems and serve as the underpinnings of many of the available TRIZ software products. TRIZ is such a rich process and tool kit that its applications and use generate different approaches around the same overall algorithm. I have found in my own problem solving and TRIZ training career over the past 10 years that only a handful of problems truly require this level of sophistication and that the basic TRIZ tools are more than sufficient and easily understood. If my description and examples resonate with you, then I am happy to have been of assistance. If I have encouraged you to learn more about TRIZ from whatever source, then I am just as appreciative.

A Tribute to A Man I Never Met

Genrikh Altshuller was a brilliant young naval engineer, inventor, and patent examiner for the Russian navy and performed the same function in the former Soviet Union that a US patent examiner does. Though the political structure of the Russian patent office was different than that in the West, and shrouded in secrecy, the principles by which it operated were very similar. The application process, followed by frequent rejection and back and forth discussion, and then followed (maybe) by an issued patent, was the same. The fact that the Russian government owned the patent (as opposed to the inventor or the company for whom the inventor worked) did not change the fundamental process. Instead of focusing on the nitty-gritty details of the patents and how they related to other similar patents in the same field, Altshuller did something that no one else had ever done before. He analyzed patents across many fields, not just his own field of naval engineering. This is exactly the opposite of what most patent examiners do. They tend to focus on the state of the art within their particular field of focus.

Altshuller began to see that an invention in one field, if generalized, was very similar to inventions in another "parallel universe" field. An example would be the recognition that the principle behind an electronic capacitor is exactly the same as that used to split diamond chunks into usable diamond dust for industrial high precision machine tool grinding ("store energy and later release it"). Why would either of these inventions be given a patent if knowledge of the other already existed? Is there a patent on holding your temper? Aren't all of these examples of the same principle, "storing energy and suddenly releasing it"? The inventive principle is exactly the same. If one looks at the patent literature 7+ million patents later, we find that 40 generalized inventive principles are continuously reused over and over again across virtually all areas of technology (and business, it turns out). The business principle of upward integration to capture more value (as done by a number of major corporations through acquisitions) is no different than integrating the paint bucket into the handle of a stick by Black and Decker. We can also think of this general concept as the way we use algebra rather than trial and error to solve quadratic equations. We don't guess at the values of "x" for *any* quadratic equation. We just calculate it because it is a subset of a general class of equations we characterize as $ax^2 + bx + c = 0$. Can you imagine what life would be like if we were still solving mathematical equations by trial and error? Why do we assume that creativity is something mysterious that cannot be generalized? I suspect that there is an incentive to many to keep it that way. Attitude, situations, and social settings can affect our problem solving capabilities, but nowhere near to the degree that many consultants would have us believe.

One day, Altshuller will receive the degree of global recognition accorded to people like Thomas Edison (the genius of trial and error experimentation), Osborne Parnes (the man behind brainstorming and the Creative Problem Solving process), and Edward DeBono (the originator of organized psychological processes such as Six Hats™ and Lateral Thinking™). Altshuller brought science and structure to all that those prior to him had done and had not recognized because their approach was not scientific. Altshuller figured out how to analyze and organize previous inventive principles in such a way as to nearly eliminate the need for trial and error experimentation in figuring out how to solve a problem. He died of Parkinson's disease in his early 70s in St. Petersburg, Russia, in 1997 without having profited to any significant degree from his work. I walked into a dining room in the United States filled with his Russian TRIZ masters the day his passing was announced. I drank vodka with them.

Disclaimer

There are instances where we have been unable to trace or contact the copyright holder. If notified the publisher will be pleased to rectify any errors or omissions at the earliest opportunity.

Contents

Part I
The Psychology of Innovation:
Attitude Adjustment

Your attitude, when you start a personal innovation journey or try to implement one inside an organization, is critical. You must ask about your attitude toward newness, your attitude toward threats to your business, and your attitude toward thinking about things and processes in different ways. We will be discussing a process and tool kit that will probably require a significant adjustment in your normal thoughts and processes that you may be using in your innovation efforts. What I will share with you, especially in the first two chapters, may be difficult to deal with but is critical to the success of your innovation journey. The concepts discussed may strike at the heart of how you currently view things. I urge you to think about them seriously and accept them as you read further in the book.

Chapter 1
Attitude Adjustment, Jargon, and Acronyms

There could be several reasons you are reading this book. The first is that you are beginning to use TRIZ and are reading several publications to get a better understanding, from varying viewpoints, of what TRIZ is and how to use it. Mine is only one perspective on this and I encourage you to read others' as well. A second is that you may have read something else I have written on TRIZ or on the general subject of innovation and find my perspective useful. A third is that you have been exposed to TRIZ in some way and find it confusing or complicated. I hope this book assists you to achieve a better understanding of this valuable and unique process. TRIZ is a very rich tool kit within an overall algorithm, similar to mathematics, and it is important to know how to use the various pieces. It is not necessary to use all the pieces all of the time any more than it is necessary to use algebra to add and subtract simple numbers. We learn to add and subtract before we learn to multiply or divide because we need to know these processes to use the others. Lastly, you might have experience with other creativity tools that have their basis in psychology as opposed to science and found them lacking in value or productivity in many challenging situations. These types of tools include simple brainstorming, structured brainstorming processes such as Creative Problem Solving, mind mapping, Lateral Thinking,® and Six Thinking Hats®. Most of these types of processes are relatively easy to learn, but lack the depth of a science-based approach to creativity and innovation that can be used for difficult or complicated problems. Any process that helps to change our perspective on problems is valuable. The difference between TRIZ and these other tools is that TRIZ provides a serious, science-based structure to accomplish this. It is possible to combine parts of the TRIZ tool kit with these other processes and ways to do this with some of these processes will be discussed later in Chap. 16.

Before we delve into the details of TRIZ, it's important that you are able to make several major psychological and mental adjustments in your thinking about creativity and innovation. The first is to get rid of the term "optimization" in your vocabulary. This is a fine word for other pursuits and challenges but not for innovation. Optimization is the identification of the minimum pain point between two undesired approaches. It is not innovation. The second is to believe that breakthrough innovation is in fact a learnable science and not some special "gift" with which a few

J. Hipple, *The Ideal Result: What It Is and How to Achieve It*,
DOI 10.1007/978-1-4614-3707-9_1, © Springer Science+Business Media New York 2012

special individuals are endowed. You have to believe that this type of talent is a teachable skill. Another thing we need to do is get rid of the fundamental belief that the only way to get a truly unique, creative idea is to generate hundreds of "uncreative" ideas. In other words, that we need quantity to generate quality. There are no psychologically based creativity tools that I am aware of that don't emphasize the generation of the maximum number of ideas possible. Have you ever asked yourself why this is so? My theory is that this is because not enough time is spent defining the problem that needs to be solved. I once participated in a brainstorming training session where the objective was to find as many new uses as possible for a balloon. Now this is a great "warm up" exercise to get everyone thinking, but it begs the question, "Why are we trying to find new uses for a balloon?" Is it because our factory losts a major customer and needs another one? Is it because the use of balloons has been outlawed for certain types of birthday parties? Is it because we have new noise regulations on balloon popping? Is it because there has been a steady decline in peoples' abilities to blow up balloons? I was told by the facilitator that the record number of ideas generated by one group of people doing this exercise in 5 min was 37. That's very interesting, but not terribly relevant. Without defining the problem we are trying to solve, we just waste a lot of time (but maybe have fun) generating and then sorting through many silly and useless ideas.

Last and most importantly, we need to set aside our egos. All of us want to believe we have special problems that no one else has ever solved before. The longer we work on solving a problem, the more important this becomes because if someone shows us an answer to a problem we have that came from reading a 10-year-old publication, it may appear that we have wasted a lot of time and money reinventing the wheel. I have given many TRIZ talks which afterwards have generated a comment like, "that's really fascinating and I'll bet it applies in many situations, but not here. We have special problems…" At that point I know that I have failed in getting across the uniqueness of TRIZ. Ego is a very strong behavior pillar. It gets in the way of business relationships, acquisitions, and partnerships. It gets in the way of successful marriages and product introductions. It gets in the way of understanding others. And most importantly for our current purpose, it gets in the way of breakthrough problem solving and innovation.

We will discuss the barriers of optimization and ego specifically in the next two chapters. These barriers will permeate all of the other materials in the book. So let's start the journey, get rid of our egos, stop optimizing, and learn from all the inventors of the world.

Chapter 2
Optimization: The Enemy of Innovation

How many times have you seen a graph like the one in Fig. 2.1?

What does it imply to you? You can't have it all, it says. These thoughts occur to you even without labeling either axis or describing the system we are talking about. We see this kind of graph so often that we don't realize that there are at least two lines on this graph condensed into one curve. There is some function, attribute, or characteristic that is improving, while another one is getting worse and the graph in Fig. 2.1 is the net sum of the two, showing the "optimal" point in the system. In my days as a chemical engineer, and in the training I do for the American Institute of Chemical Engineers, this curve is commonplace in training and analysis when describing many chemical process unit operations. The more insulation you put on a pipeline, the less heat you lose, but the cost of the insulation (as well as its installation and maintenance cost) increases. There is an "optimum" insulation thickness that is a function of the cost of energy and the cost of insulation and labor at a given point in time. When a process plant is built, the design engineer, with possible input from long range planning thinking about the price of energy, calculates the cost of the energy required to keep a process pipe at a certain temperature versus the cost of the thickness of the insulation materials (or the tracing system if steam or hot oil is used). This is exactly the same calculation and thought that goes into the decision about how much insulation is installed into a home. The cost of the insulation system is prorated over the projected life of the process plant, taxes and depreciation are figured in, and a graph such as the one is produced (Fig. 2.2).

Chemical and mechanical engineers who specialize in this area have numerous computer programs and algorithms which can optimize this design, taking into account insulation costs and the cost of energy in different forms (steam, electric, hot oil). Every once in a while, possibly triggered by a significant increase in the cost of energy, a review of these decisions is made and the possibility of changing the type or thickness of the insulation is made, using the same kind of cost–benefit trade off analysis. In this discussion, we have *assumed* that we needed the insulation. It's easy to fall into that trap, isn't it? We have a system that we have had for a long time. It's not perfect and we slowly but surely try to optimize it based on all

J. Hipple, *The Ideal Result: What It Is and How to Achieve It*,
DOI 10.1007/978-1-4614-3707-9_2, © Springer Science+Business Media New York 2012

Fig. 2.1 Classical optimization
curve

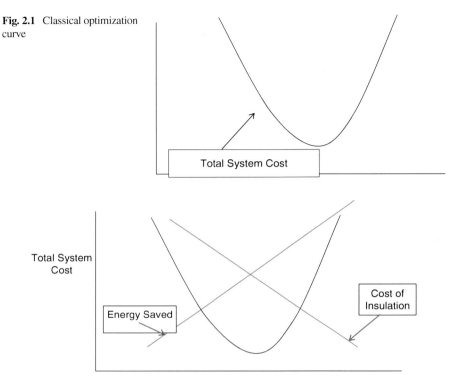

Fig. 2.2 Optimization of insulation

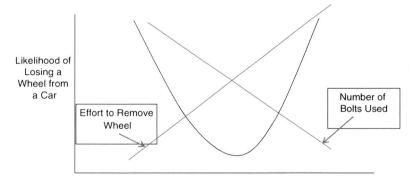

Fig. 2.3 Wheel loss optimization

kinds of external conditions. Optimizing like this means balancing two negative
things, which surely does not produce an Ideal Result.

Let's take a look at another similar example that everyone is familiar with—the
tire and wheel on a car. A graph similar to the heat transfer optimization graph
would look like the one in Fig. 2.3.

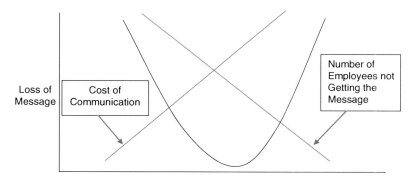

Fig. 2.4 Optimization of organizational communication

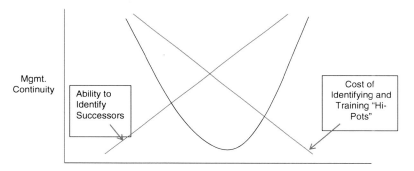

Fig. 2.5 Optimization of organizational succession planning

Over the years, the tire, wheel, and auto companies have "optimized" the choice of the number of bolts to minimize the pain while minimizing the chance the wheel will fall off. Is this really an Ideal Result? What if we didn't have to tighten any bolts at all? (Hold that thought!).

What does a graph summarizing internal organizational communication as shown in Fig. 2.4?

We can't afford to tell everyone everything, can we? So we make a judgment call about how much information is needed by whom and rely on our managers to communicate "down the line" as they see fit, potentially filtering the information from getting to everyone who might need or be interested in it. We also make choices about which mechanisms we use to communicate—written materials, bulletin boards, Emails, and other electronic forms of communication (Fig. 2.4).

What about succession planning in an organization? The graph might be the one in Fig. 2.5:

"Hi-pots" is a term used in many companies to designate young "up and comers," thought to be candidates for senior management positions.

What are the other business examples of this type of analysis? As an organization becomes larger, communication becomes an issue. Many companies have a

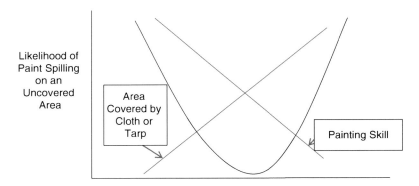

Fig. 2.6 Optimization of floor covering while painting

deliberate policy of not allowing site sizes to grow beyond a certain level. This maximizes the efficiency of communication within a site and starts to inhibit communication between sites. One problem gets easier and another becomes more of a challenge. We have a tradeoff that we think must be managed and "optimized."

Consider communication with customers. How much is too much? How do we decide how often to touch base with a current customer and "see how they're doing?" Do we wait for a complaint? How much travel money needs to be spent communicating? Are there alternatives? Is the optimization the same for all customers? Is the solution the same? What affects these decisions? This is analogous to the decision about how much insulation is needed when the cost of energy changes. The thought process is the same.

Let's look at painting. When you are upon a ladder with a roller pan and paint roller, you don't want to get paint on the floor. You make a judgment call about how clumsy you are and decide how much and how thick and what type of drop cloth you are going to use. Here's the graph that went through your head without realizing it (Fig. 2.6).

What if we didn't have to worry about the paint spilling and did not need a tarp or cloth? (Hold that thought too!).

Back to chemical engineering. In a distillation column, we trade off the number of trays and the reflux ration in the column as in Fig. 2.7.

The more trays we put in a distillation column, the less reflux (material returned to the column) is needed and the lower the energy cost to reboil the liquid up into the column. The capital cost of the column now increases. If we shorten the column, we need more energy to produce the same separation. Large amounts of time and money are spent on this optimization in the oil and petrochemical area to minimize the total cost of producing gasoline, heating oil, and a myriad of petrochemicals such as ethylene, propylene, and styrene—the backbone materials of our hydrocarbon, fuel, and plastics industries.

If we consider the business analogy to this example, consider the pricing of chemicals and materials versus their purity. Some people will pay more for 99.99 % purity, others will not. For them, 99 % may be just fine. What does this optimization

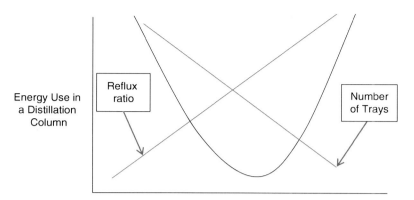

Fig. 2.7 Optimizing a distillation column

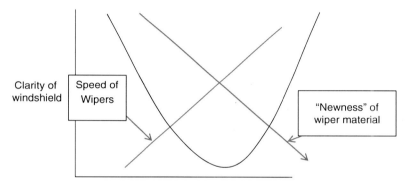

Fig. 2.8 Optimization of windshield wiper speed

curve look like? If we put our TRIZ hat on for a short while, we might begin to ask how we might change how a lower purity material can match the performance of a higher purity material. Why do we have to move back and forth along the curve? How can we move the curve?

Consider the windshield wiper speed in your car during a rainstorm (Fig. 2.8).

The more frequently we sweep the windshield, the more rain is removed and the clearer our vision. However, this wears the blades and leads to their earlier replacement. What if we didn't need wipers at all? Have you seen the Rain-X® product?

We have the issue of protecting children from getting into medicine bottles, while at the same time making the bottles easy to open for people with arthritis (Fig. 2.9).

This is an example of a design contradiction that has significant social implications. No one wants to a see a small child poisoned by medication, but at the same time we don't want to see the elderly not take their medicine because they cannot open the bottle. Why do we need the cap? Why do we need the bottle? Is swallowing a pill the only possible way of delivering medication? Is that the only way to

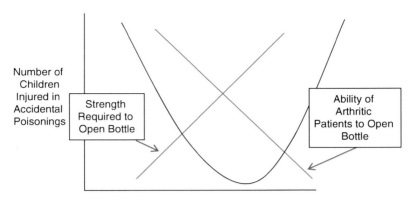

Fig. 2.9 Optimization of pill bottle design

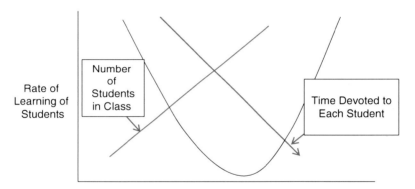

Fig. 2.10 Classroom optimization

dispense medicines? What other ways are there to dispense medicines in a way that prevents any accidental exposure to young children? Is there a way to prevent children from opening the bottle? What is fundamentally different about children's physical traits and adult traits that eliminate the potential danger to a child opening a bottle? Let's not minimize the potential danger—let's *eliminate* it!

Other examples we could plot might be the optimization of chemical reaction, temperature balancing rate, and yield for an exothermic chemical reaction or the manner in which we administer a merit increase or bonus program (how do we motivate our best people without demotivating our "average" performers?).

Let's look at the classical contradiction graph that is constantly discussed by teachers, parents, teacher unions, and boards of education (Fig. 2.10).

Many parents, teachers, and teachers unions believe that classroom size is a key determinant in learning. This may be true in some circumstances, but there are many examples where this is not true and some other factor may be more important. If we plot performance against only one variable, we may not see the whole picture.

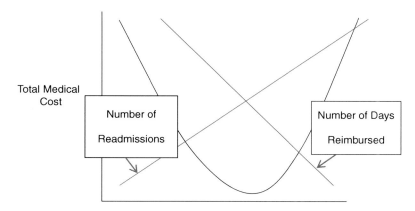

Fig. 2.11   Insurance optimization

Table 2.1   Optimization table

This gets better	This gets worse	Optimization control variable(s)

Might nutrition, family life, teacher competence, or learning materials have an equal effect? We don't know and a graph as seen in Fig. 2.10 can deceive us.

Let's take a look at one last optimization graph, representing how a medical insurance company decides how long to reimburse the patient and hospital for planned surgery or pregnancy (Fig. 2.11)?

If we send mothers with newborns home in less than 48 h, we minimize the immediate cost of the stay and reimbursement, but run the risk of a significant readmission cost should a complication arise. Is "optimizing" the stay time the best way to handle this?

I could go on and on with dozens of additional examples, but you get the point. You can do the same thing for your technology, process, social, or organizational issue. Make a list of the compromising issues you face (and make sure you include both business and organizational issues in the list!) (Table 2.1).

Keep these thoughts in mind when we discuss contradictions in Chap. 10.

If you appreciate graphical representations, take these tradeoffs and make graphs similar to the ones I showed earlier. A template is included in the appendix for your use (See Fig. 2.12).

We tend to think of the world as a choice between good and bad, resulting in a decision to minimize the pain as determined by some arbitrary definition. In areas of morality and religion, the concepts of good and bad may be perfectly acceptable, but in the real world of designing and operating equipment and processes or in the world of organizational management, things are never this black and white. The "optimum" may also be different for different stakeholders in the above cases. Surely the maker of wheel bolts does not see the wheel situation the same way as

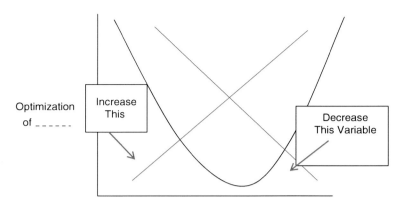

Fig. 2.12 Optimization template

the auto driver who must remove the wheel. The supplier of paint tarps does not see things the same way as the homeowner who would like to find a way not to use tarps at all. The maker of distillation column trays does not see things the same way as an external utility supplying the steam. In our current discussions about improving health care and its delivery, the trial lawyers, insurance companies, hospital administrators, sick patients, doctors, and nurses all have different visions of the "optimum" goal and the way to achieve it.

In thinking about our current health care debate in the USA, do the patient, the nurse, the doctor, the hospital, and the medical insurance company look at things the same way? Certainly not. What would their graphs look like? What variables might these other professions plot on such a graph?

The point here is that the optimization of a system and the Ideal Result (which we will discuss in a later chapter) is not necessarily the same as viewed by all stakeholders. The tools of TRIZ do not provide a direct answer to this choice, but some of its tools can point the direction we should look to make the best choice. We'll discuss this in more detail later.

Now, having thought about "optimizing" for a while, ask yourself—are you happy with this approach? You may have minimized the pain, but it's still there. We may have moved the pain to somewhere else in the system. It's like taking a sleeping pill every night to compensate for too much caffeine or paying some of your deductible housing expenses to minimize taxes this year—which increases the taxes due next year. When you do this, are you really "innovating" or just optimizing and compromising? You are working within the constraints of the system _as it exists!_ Now of course, if a product or system has not been analyzed and optimized in a long time, the result may appear to be an "innovation" and is certainly worth doing and may have significant cost savings, but as long as we are trading off, I submit we are only improving and not innovating. We are operating and analyzing within the framework of what we know and already do. There is nothing wrong with going after the low-hanging fruit, but do not let this activity distract you from attacking longer term, more permanent solutions.

This brings us to a couple of the fundamental principles within the TRIZ tool kit and algorithm. The first of these is the concept of the Ideal Result. Those of you who are into optimizing, try to imagine what life would be like if the issue or parameter that needs to be optimized simply disappeared as an issue of concern. (By the way, this is no problem for a child with some imagination, as they do not know all the reasons you can't do something!) Imagine there is no reason to do succession planning, no need to constantly update the computer program that calculates the optimum reflux ration, and no need to worry about the wheel falling off the car. Don't you feel better already? Outsourcing is one thing that corporations do to get rid of this pain and hassle, especially if they believe their competence in a particular area is weak. But again, we still have only minimized the pain and put it out of sight.

Let's relook at some of the previous examples. In the first case, the pipe insulation—what if there was no insulation thickness to optimize? Now I don't mean that the material inside the pipe freezes or boils and we just live with that result and call the necessary maintenance people when we need them. I mean that the material in the pipe and other components in the system *other* than insulation *control their own temperature.* Or maybe we use a fluid whose physical properties make it less likely to freeze or boil. Just think about that while we look at case 2—the wheel. What if there were no wheel bolts? There's nothing to optimize because the bolts are gone and the wheel still stays on. In the communication example, what if we had no communication meetings to plan, organize, and optimize? How much money would that save? In succession planning, what if we spent no time analyzing and vetting future leaders? What if they just appeared when we needed them? A lot of time and money would be saved.

In the painting example, what if we had no pan that might spill, no roller that might sloop and leak? We wouldn't need a big tarp! In the distillation column, we are balancing capital and energy costs. What if we did not have to separate the materials in the first place? No steam, no column. Why can't the car window stay clean without wiper blades? Then there's nothing to adjust. The window cleans *itself.* Finally, in the case of the pill bottle, if the lid goes away, there's no force or system to optimize. How can pills be stored and delivered for adults in a way that doesn't require a bottle the size kids want to play with? What if the cap and bottle geometry were such that it was impossible for a child to open, but easy for someone with arthritis to open?

Do you see what we are starting to do here? We are getting out of the optimization mind set and getting to a type of root cause thinking, but far beyond normal root cause thinking. If we *get rid of the problem* (the thing or variable we are trying to optimize), then there's nothing to optimize—there is no problem we need to solve. How much money is that worth? The problem and the system become simpler because we eliminate something. *Anything* that is eliminated from a system makes it simpler, costs less, requires less maintenance and attention, and improves reliability. It's easy to conceptually eliminate a part of the system. The challenge is to follow that thought with a serious, methodical methodology that figures out how to eliminate a part or system and still achieve the result or function that it was originally put there to do. That's what the rest of this book is all about. But I want to emphasize that without the desire and capability to make this fundamental mind shift, the rest of what you will read will probably be interesting (like a fairy tale you enjoy) but not relevant or useful. So spend some serious time reorienting your brain and thinking about the Ideal Result. Stop optimizing, trading off, and minimizing pain. Let's get rid of the pain and stop taking pain relievers.

Exercises

1. Take a product or service you provide or use and seriously think about the tradeoffs which have occurred, over time, in performance, operability, user friendliness, or other key parameters. Make a simple, qualitative graph if you can. How have these tradeoff decisions been made? On what basis? Why? What has been the cost of the "fix" versus the perceived cost savings? Has the root cause of the problem ever been identified? Eliminated? Why or why not?
2. What is the key variable that is used to control the optimization? If you don't know, why not? Is there more than one? What is its function? Is that function *really* needed? Why? If the function is really needed, how else could it be achieved?
3. Has your optimization graph changed over time? Why? What caused the change? If you don't know, ask yourself why and find out.
4. What is your rationale for "optimizing?" Does it really make sense? Why? On what basis?
5. Who drew the optimization graph? How many views of this graph might there be? Do others use the same "y" and "x" variables? Why or why not? Is there a variable that some see as important and other don't? Why? What are the implications of different optimization graphs?

Use the optimization template (Fig. 2.12 and also in the appendix) to assist you in doing this.

Chapter 3
Parallel Universes

The second step in changing your mind set about innovation (after we start thinking about breakthrough as opposed to "optimizing") is to get rid of the idea that the problem you are trying to solve is unique and special, and that no one has ever encountered a similar problem. Now I realize that there are truly challenging problems out there, but in my years of TRIZ work I have seldom seen a problem that, in a general sense, has not already been solved in a parallel universe. I don't mean that an engineering drawing for the solution exists. I mean the general concept of the solution exists. This general concept then needs to be applied to the specific problem at hand. This is the same observation that Altshuller made when he started to review the patent literature. The exception I will make to this assertion is the discovery of new science or a new scientific law or principle.

Let's start with a very recent example. Last year, on a flight out of Houston, I glanced at the airline magazine in the seat back pocket. The back cover caught my eye and I noticed an advertisement about a joint conference in Houston, involving a joint meeting between the oil and gas industry and the heart medical community ("industry") in the Houston area. The title of this conference was "Pumps, Pipes, and Valves" (see website, http://www.pumpsandpipes.com). This conference between these two technology areas has been going on for only a few years and is in the process of expanding to overseas locations. My recollection is that these two major industries have been centered in Houston for over 50 years and yet it took 47 years for them to "discover" each other. They both use "pumps" (one biological and one mechanical). They both are concerned about pressure drop and lines plugging and about mechanisms and means of reducing friction and pressure drop. Yes, the scales are different, but as far as I know from my chemical engineering training, friction factor, Reynolds number, Bernoulli equations, and pressure drop calculations are independent of scale. It's the nature of the fluids, "pipes," surface properties, fluid viscosities, friction factors, etc. that really matter. I have to wonder about the possible impact if these two industries and their specialists had started talking seriously decades ago. What would the state of technology development be in both? How many fewer heart attacks? How much more efficient would oil and gas recovery be?

J. Hipple, *The Ideal Result: What It Is and How to Achieve It*,
DOI 10.1007/978-1-4614-3707-9_3, © Springer Science+Business Media New York 2012

Another classic example that we can relate to is the use of stored energy. Ask yourself what is the difference between these inventions and concepts, whose patents (in some cases) are separated by decades:

1. Use of pressure to cause a crack in a diamond, air entering the diamond, and then suddenly releasing the pressure to produce diamond dust
2. Use of pressure to cause a stress crack in a pepper, allowing air to get inside, and then releasing the pressure to explosively remove the stem in the pepper, allowing it to be processed into salads in a bulk fashion
3. The use of pressure to penetrate rice granules and then suddenly releasing the pressure to make a puffed rice product
4. The use of dissolved cryogenic gas to penetrate paper bundles in an aqueous solution, and then releasing the pressure to "explode" the paper into pulp particles
5. Dehusking sunflower seeds
6. The concept behind an electrical capacitor
7. The concept of a battery
8. How a dam works to "store" power
9. Holding your temper
10. Slow release fertilizers

In all of these cases, we are storing different forms of energy and then releasing that energy at a later time. Why should there be multiple patents that use this same principle? Later on, when we discuss the TRIZ contradiction table, we'll see that in reviewing millions of patents, which is where we document inventions, we find a limited number of inventive principles that we constantly reuse across a myriad of different industries and technologies. The specific jargon and terminology we use in many situations makes us *think* our problem is special and unique. It rarely is.

Let's consider some other examples to get us thinking about this point. Suppose you were trying to deal with the problem of improving the design of an air traffic control display. One of the issues here is that government regulatory agencies in the USA, such as the FAA, want an air traffic controller to have more and more information on the screen in order to assist in making the correct decisions. The problem is that the more information you have, the more difficult it is to analyze that information and make critical decisions and provide timely input to a pilot. You can see the outline of what we call in the TRIZ world a contradiction. In other words, we would like the information to *be there* for one reason and *not be there* for another. We'll discuss contradictions in much more detail in a later chapter, but for now just think about where else a problem of this nature exists. Who else has to deal with an abundance of information and react and make decisions regarding it rapidly?

Consider:

1. Chemical plant process control displays
2. Surgical operating room displays
3. Video games
4. Automobile dashboard displays

What others can you think of?

Recently, we see some signs that these parallel universes are becoming aware of each other, not too differently than what we saw in the oil and gas well and heart artery parallels. In an article published in Medpage Today (2/19/2007), it was reported that "surgeons who played video games least 3 h per week, were 27 % faster with 37 % fewer errors is simulated laparoscopic surgeries."

One of the self-created barriers in recognizing the general, as opposed to specific nature of the problem we are facing, is the jargon and special language and terminology we use to describe our problem. Using special words supports our belief that our problem is special and unique. For example, have you ever heard or used the word "defalcation?" Do you know what it means? I first heard this word in a problem-solving session with a major Canadian bank.

Take a look at this website: http://en.wikipedia.org/wiki/Defalcation.

You'll see that defalcation refers simply to the misappropriation of funds, usually in a financial transaction. Now type the word "defalcation" into your Web browser. How many hits do you get? On the day I did this I got between 60,000 and 350,000 hits depending upon which search engine used. What happens if we enter "fraud" into the search box? We get between 40 and 310 _million_ hits or about two orders of magnitude more information to analyze that relates to the problem. Using "substitution of one thing for another" will produce the same kind of results. Where would you rather look for a new idea in this area? Why do bankers use a special word that automatically limits the boundary of thought? It does make the problem seem "special," doesn't it? Don't we like to feel special? Don't we like to think our problems are special? We learn from TRIZ and a study of the inventive principles documented in the patent literature that _there are virtually no special problems!_

Where else do people try to "defraud" customers, clients, or organizations? Art sales, antique shows? I wonder if a bank "defalcation" expert has ever attended an antique show or interviewed people who make a living out of selling what others _think_ are antiques.

In another medical area, heart stents are a topic of major concern. They are inserted into the arteries of people, whose arterial walls have collapsed, creating blockages to blood flow. The problem is that, after a period of time, tissue starts to grow around the stent, creating a new blockage. The materials used in these stents are special plastics and polymers that are derivatives of convention plastics used in many industrial and household applications. A major medical device company was considering a method to allow the stents to slowly dissolve back into the blood stream, allowing the artery to stay open. Where else do we want plastics to slowly dissolve? Landfills! See http://blogs.wsj.com/health/2009/03/13/abbotts-search-for-a-disappearing-stent/. A question you might ask yourself, if you were a research manager at a heart stent supplier, what would your response be to a request from someone on your technical staff to go to a conference or convention focused on plastics degradation to lengthen the life of landfills? Would you encourage them and pay for the trip, or ridicule (maybe fire?) them for asking to spend money on such a stupid trip?

Researchers at Emory University and Georgia Tech, in a similar joint effort, have begun to study the fluid behavior in arteries and how it affects plaque buildup (http:// shared.web.emory.edu/whsc/news/releases/2011/08/predicting-perilous-plaque-via-fluid-dynamics-in-coronary-arteries.html). They have also made the observation that figuring out where plaque will "deposit" is the same as trying to figure out where sediment in a river will deposit. Shear stress is the same phenomenon whether it occurs in a river, a pipeline, or an artery. Reynolds numbers and settling velocity concepts do not change with the system.

Do you remember the Proctor and Gamble product known as Olestra™? This was "zero fat" oil that was basically nondigestible, reducing the net calorie content of snack items such as potato chips. One of the unfortunate side effects of this product, due to its lubricity, was anal leakage, limiting its commercial success. Where is this negative aspect viewed as a positive? Where else do people care about lubricity and lowered viscosity? Two such industries include paints and machine lubrication.

Forbes (http://www.forbes.com/forbes/2009/0824/outfront-olestra-health-paint-it-fat_print.html) and Wikipedia (http://en.wikipedia.org/wiki/Olestra) describe how P&G is now exploring the use of these properties of Olestra™ in these parallel universes. It is possible that Olestra™ could replace millions of gallons of volatile, organic chemicals used in solvent-based paints. This work is occurring a decade after Olestra's introduction into the food industry. Several questions to ask yourself at this point. What if P&G was a coatings company instead of a consumer products company? What if a paint company had discovered the chemistry behind Olestra™? Would they have looked at applications in food? How much money could have been made by Proctor and Gamble if it had looked at the paint market early on, collaborated with a paint or coatings company, and then manufactured the product and licensed its use in a parallel universe?

Exercises

1. Try describing your problem to your 10-year-old child. Do you use the same language? What words _did_ you use? Now use these words in your search engine. What new perspectives and ideas do you see?
2. Have you ever had to answer the question (from your kids, your wife, your relatives?)—What do you do for a living? Were the words the same as you used in your discussions with your clients or customers? Use the more general words in your search related to your problems and challenges.
3. After you have gone through a couple of these mental exercises, conduct a Web search for the "association" related to the more general terminology. You will find that there is an association for _anything!_ For example, all of the previously cited cases related to the interface between the human eye and a display screen are of interest to the Human Factors and Ergonomics Society (http://www.hfes.org).

Anyone concerned with how humans interact with graphical display information would benefit from learning from all the other people who have the same generic problem. If you were to attend an HFES meeting, as I have, you would discover a wonderful breadth of disciplines in its attendees. Going back to our bank fraud example, did you know that there are associations specifically concerned with this area? See http://www.acfe.com/ and http://www.nafraud.com/.

Before we move on to the more formal aspects of TRIZ problem solving, it is critical that the concepts in these first two chapters have been absorbed and internalized. If you still believe that optimization is the key to breakthrough innovation and if you still believe that the problems you need to solve exist nowhere else in the universe, go back and reread the first two chapters. If you cannot accept these two basic premises, you will find the rest of this book interesting but probably not relevant or useful, or you may say that this is very useful stuff for everyone except yourself, your company, and your products and businesses. If so, I can understand, but will tell you with a great deal of confidence that someday you will be surprised that someone else, in an area unknown to you, will have solved your problem and you will pay a licensing fee to use the technology.

Part II
TRIZ Thinking
and Problem Solving Tools

In this section, we will cover the basic TRIZ problem solving and analytical tools. They can be used independently or collectively. They can be used as a separate process or used alongside other creativity processes that you may be using. It is recommended that for maximum value they be used collectively in an algorithmic type fashion. We will build this algorithm as we discuss the various tools. The fact that creating new breakthrough ideas and solutions to problems can be a defined process will be difficult to adjust to and will not seem to be as much "fun" as other approaches you may have used in the past. You will be tempted to jump into the solution space before you have adequately defined the problem. Be patient and try the process. Think about some of the problems you may have dealt with in the past and remember the random way you may have arrived at the answer and compare that with using this process.

Chapter 4
The Ideal Result

We have now set the stage for the first tool and analysis part of the TRIZ problem-solving algorithm. Let's think back to some of the examples presented in the previous chapters. If we could eliminate the variable that we needed to optimize, our system would be much simpler and since there's one less variable or part to deal with, our system becomes more ideal. Let's now define the Ideal Result. It's when *something performs its function and does not exist*. There are no costs or negative side effects. It's as if the result we are looking for all of a sudden arrived in a gift package from an anonymous donor. This is a difficult concept to grasp as most of us, while thinking this way, are already thinking about all the reasons that this cannot be achieved. Now TRIZ does not necessarily guarantee that this will be the result, but getting 80–90 % of the way there is much better than what we normally accomplish. This first "envisioning" step is absolutely critical to the rest of the process.

The "something" can be a part, a person, a piece of machinery, a maintenance procedure, a process, a message, a software package—*anything* in a system, whether it be a mechanical, communication, electronic, or personnel related. The problem is that we get focused on the "something" and not its *function.* What is it *doing? What is it accomplishing?* That's where the focus needs to be. We want the *result.* We really don't (or shouldn't) care about the "something" unless possibly we are designer or supplier of the "something." Then we have a vested interest in preserving and optimizing the "something" that may not really be needed. In fact, if you are the customer of the supplier of the "something," you would much rather not receive an invoice, correct? You'd like to get the "something" without paying for it. (Isn't that what happens with many new Internet and phone services?). In fact, if you're a forward looking company, you probably have a team trying to reduce the amount and cost of the "something" that you use. Your purchasing department does this frequently with your suppliers, irritating them to no end. Most of the time this is simple competitive bidding to get a slightly lower cost, but you may also have a team trying to figure out how to stop using the "something" or replace it with something else. Paint trays aren't a huge business but they are a necessity for normal painting. A company like Stanley Black&Decker knows that not using them would

J. Hipple, *The Ideal Result: What It Is and How to Achieve It*,
DOI 10.1007/978-1-4614-3707-9_4, © Springer Science+Business Media New York 2012

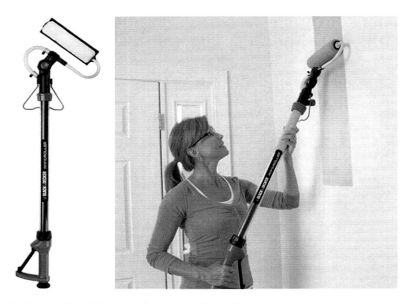

Fig. 4.1   Stanley Black&Decker pivoting RapidRoller™ system

be a plus to its average consumer. How concerned are they about the profits of a company supplying paint trays? Or the supplier of the aluminum needed to make the trays? The manufacturer of the equipment used to form the tray? Stanley Black&Decker is a major supplier of household tools and products to help the homeowner accomplish jobs. But they know we still need the paint tray's *function.* And what is that? It's to "hold paint between roller applications." But is that their real function? It's easy to get trapped in the jargon of what we are doing. The function is not to hold paint—it is to provide *volume* to hold the paint. This is a subtle but very important distinction. Now the tray is one source of volume. Might there be another? What about the volume *inside* the handle? If we think of it as a paint "stick" or "rod," we immediately think of something solid. But why does it have to be solid? That extra wood just costs extra money. If the stick is hollow, it is not only cheaper (uses less metal or wood) but also provides volume that could be used to hold paint. So the first step in getting started here is to clearly state the Ideal Result: *something performs its function and does not exist.* In the application of this mental process, we not only simplify the system (get rid of something) but also we save money. Why do we want to pay for wood or metal volume that is not performing a useful function? Stanley Black&Decker's Pivoting RapidRoller™ system uses the hollowed out volume of the roller handle to hold paint and feed it to a roller. The stick is connected to the bucket and the paint drawn up into the handle. The ladder and paint pan are eliminated. A smaller drop cloth is required (Fig. 4.1).

Now stop and ask yourself—why couldn't this innovative product have been developed decades before? Is there any new law of physics that needs to be discovered to "invent" this product? Hollow tubes, artificially created vacuum, mechanical seals, and suction pipes have been known for decades. Has anyone who has painted anything ever been excited about cleaning up paint spills, laying down tarps, or climbing ladders? Did we need a marketing research department to tell us that people would rather not have to do any of these things? It is likely that a "normal" marketing research interview is showing the person being interviewed the elements of the currently used painting system (paint can, ladder, spill protective sheet, paint pan, etc.) and asking the painter what features they might want to improve. Suggestions like these might come back:

1. Make the drop cloth more absorbent
2. Redesign the roller so it won't drip as much and hold more
3. Make the ladder more adjustable

You've seen products over the past decade that responded to these needs, including more absorbent cloths, "gel" paints, as well as ladder systems that will match any painting situation. What do all of them have in common? They all add cost and complexity. The Stanley Black&Decker product meets many of those same needs more simply and more economically.

In my experience with clients in TRIZ problem-solving sessions, solutions generated by TRIZ always look so simple in hindsight that a group is amazed and frequently embarrassed. After the fact, TRIZ is often not given proper credit because the solution, in hindsight, looks so simple.

Let's consider the distillation column example discussed earlier. We were trying to *optimize* its performance in a number of ways to minimize energy and capital costs. There was no discussion of *eliminating* it. What's its *function*? It is to separate two or more liquids from each other by taking advantage of their volatility differences. Hopefully, a good chemical engineering team and the chemists they work with have considered other methods of separation and reached the conclusion to use distillation based on cost analysis, temperature sensitivities of the materials, pressure drop across the column, etc. Did they consider all the other ways these materials might be separated? Could one be reacted away? But the more important question, from a TRIZ and Ideal Result perspective is not to optimize the separation, but to *eliminate* it. Why do we need to do the separation in the first place? Does the material really need to be separated to perform its function? How did a mixture result in the first place? Probably some other chemical step prior—possibly a reaction system that produced more than one product or a reaction run in a solvent that now has no useful purpose and needs to be recycled. How could the reaction be run without a solvent? Could it be run in gas phase with no liquids at all? We frequently assume that the purer the product, the better. Do we actually know this for a particular use? Where is the data to support the decision to run an expensive separation system? Maybe the focus of our work should be upstream in the reaction system to make a pure product which does not require purification. Now if you're in the business of selling distillation column internals and design skills to a client, what

Fig. 4.2 Typical
prescription bottle

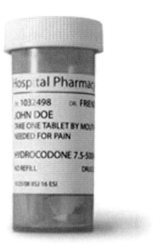

incentive do you have to ask these kinds of questions? Is your definition of the Ideal
Result the same as your potential customers?

Let's look at one of the organizational challenges mentioned previously—
communicating across a large organization. I can remember quarterly meetings in the
large companies that I have worked; being held to make sure that everyone heard the
latest news, edicts, and changes in priorities. How much time and money was spent?
What is the result that was being sought? The senior executives in the organization
wanted everyone in the organization to get the news at the same time. The Ideal Result
would be for this to occur without the time and expense of the meeting! Unless it is a
very rare occasion (an acquisition, a large downsizing, a major change in compensation programs), today we do this via Email, various electronic devices, or Intranet sites
within the company. It's the employee's responsibility to "tune in" and find out and
not complain if he don't know what is going on. If the communication is done well
and the supervisors are well informed via smaller, less expensive meetings, people get
their questions and concerns answered without having a mass meeting.

What about wiper blades? An incredible amount of money has been spent and
patents filed around the concepts of optimizing the geometry, polymer formulations,
and control systems for windshield wiper blades. Even the simple invention of variable wiper blades was the subject of an outstanding movie a while back, "Flash of
Genius." Why do we need a wiper blade? What's its *function?* Let's be careful here.
Is it to remove water from the windshield? Or is it to *keep the windshield clean?*
What's the function of a clean windshield? Do we really care about a clean windshield? Isn't what we really care about the ability to see the traffic, oncoming vehicles,
cars in front of us, people, signs, and maybe emergency vehicles? How could the *function* of the windshield be performed *without its existence?* Then we wouldn't have to
clean it, would we? Could we do this with some kind of virtual reality technology?

Let's consider a typical prescription pill bottle as seen in (Fig. 4.2).

What do you see? The name of the pharmacy, phone number and address, the
prescribing doctor, dosage, name of the medicine, etc. While everyone else is

optimizing the label, the color, the size, and other physical aspects, let's look at its *function*. What is it? To hold pills? Prevent contamination from an external source? Prevent access by children to adult medicines? Prevent degradation from sunlight? The Ideal Result would be to have these functions performed without the bottle. This is what we need to think about—not optimizing the bottle design, though that may be a perfectly fine short range group of projects for people who make their livelihood supplying plastic, labels, and inks. If the *function* of the bottle is to hold pills, how else could we do this? That begs the question of what is the *function* of the pills. We only have the bottle problem because we have pills. What's in the pills? A drug. What do we do with the pills? We usually swallow them with a fluid (water, juice, milk, etc.) to get them into our digestive tract. Is this the only way to do this? We're getting ahead of ourselves a bit here, but don't we swallow food? What can't a drug be put into food? Think about Listerine® breath strips and how their utility has spread to areas beyond the original concept of breathe freshening. If someone figures out how to legally and ethically deliver some of the more common generic drugs, I am not sure I want to be in the pill bottle business. And how excited are you about being the producer of plastics used in the manufacture of the bottle? The manufacturer of plastic molding equipment used to make the bottles? The supplier of colorants for the compounded plastic? The supplier of the paper used in making the label and the machinery used in making the label? If you're in the pill bottle business, you need to be thinking constantly about how the *function* you provide could possibly be delivered in another way because the drug store's primary objective is *not* to buy bottles. It is to provide a way to deliver the *function* of the drugs, not the drugs, to the consumer. This is a very hard concept to get across, but the history of many inventions, both technical and commercial, testifies to its truth. If you're in the bottle business, or in the business of supplying plastics to bottle manufacturers, you are faced with a difficult decision. Do you continue to "optimize" the bottle and the materials used to make it, or do you make an acquisition or invest in a business totally unknown to you? But in order to get to this point, we had to consider the *Ideal Result* and the *function* of the bottle.

Let's look at a new prescription bottle from Target that has been on the market for several years (Fig. 4.3).

What do you see that is different? In what ways is it more "ideal" than the traditional bottle? The relative importance of what's on the label has changed dramatically. The most important thing on the label now is the *name of the medicine*. When has this *not* been the most important thing on the label? Note that the "bottle" has a colored ring around it. This is to link the prescription and the bottle to a particular user in the family. Is this fool proof? Of course not, but it's certainly better than all bottles looking alike. Each family member, registered at the pharmacy, has a designated color stored in the computer register. Is the bottle round? No, and because it isn't, the label is shaped differently, requiring less paper (sorry you paper suppliers, but the information on the label is what is important, not how much paper is used).

Look at the label. We already discussed the prominence of the medicine name in bold print and at the *top* of the label. (Look at Fig. 4.2 again and see where the medicine name is). Look in your medicine cabinet and see where it is on your prescription bottle label. Are the pharmacy names, location, and phone number more

Fig. 4.3 Target® prescription
bottle (http://en.wikipedia.org/
wiki/ClearRx)

important than this information? Do you call the pharmacy in an emergency? No!
You call 911, that simple-to-remember phone number. And what's the first question
they ask in a case of someone swallowing the wrong medicine? *What did they swallow?* Let's make this the easiest thing to find on the label!

Look at the shape of the bottle. It is u-shaped allowing a label design with less
paper that also allows safety precaution information to be stored behind it easily
instead of a separate loose piece of paper that is easily lost and forgotten. Did you
notice that the bottle is upside down? Why? Take the time to visit http://www.target.
com and read about the history of this new prescription bottle design. Why did it
take the death of a pharmacist's child to trigger this knowledge? It's a step toward a
more ideal design as opposed to "optimizing" the paper strength, the printing ink,
and the label design. But we'll come back to this later on to follow up on some of
the questions we asked earlier that might make the bottle design irrelevant. For now,
think about how you would make the bottle, or more generally, the dispensing of
prescription medicines even more ideal from these viewpoints:

1. A prescription bottle is frequently in a household where there are users of very
   young age and those with arthritis. How do we make the bottle easily accessible
   to the arthritic and *impossible* for the toddler to open?
2. How do you make sure that someone *never* takes more than the prescribed
   dosage?
3. How do you make sure that the drug is *never* taken with an interfering food? (For
   example, the enzyme in grapefruit juice destroys certain classes of prescription
   drugs)
4. How do you make sure that the drug *is* taken at the appropriate time and in the
   appropriate amounts?
5. How do you make sure the prescription is refilled as scheduled? *Especially* with
   elderly patients who may live alone?

Fig. 4.4 Michelin's Tweel™

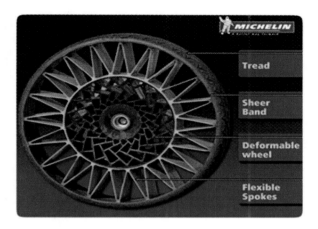

Let's go back to an earlier example, the optimization of the number of wheel bolts. What is the *function* of the bolts? To keep the wheel from falling off. How can this function be performed without bolts? Look at Michelin's Tweel™ (Fig. 4.4):

Do you see any rims? How would you like to be the rim supplier to this auto or tire manufacturer? Or for that matter, the metal supplier to the rim supplier? The company that makes the equipment that shapes the raw metal into a rim? The ore supplier to the metal producer? Are your sales going up or down? Though this tire design still uses rubber (or some other elastomer?), I suspect its properties are significantly different than what is used now—not a small percentage change from the current product. Were your salespeople asking Michelin how you could improve their tire rubber formulation or how could you help them get rid of the expense of the normal wheel? Would you see that as in your best interests? Now this example illustrates two specific techniques ("trimming" and "upward integration") for getting to the Ideal Result, which we'll discuss in more detail later, but that's not the point here. It's to get you to think about this fundamental concept of Ideal Result and the major change it might make in your business analysis. This applies to you both as a customer or as a supplier.

What about Michelin? They're in the tire business. Do you think Ford, GM, Toyota, Nissan, or Chrysler would buy tires if they didn't need to? What's their function? To allow the car to roll? Isolate the car from potholes? To help the car stay steady during heavy rains? Are conventional tires the only means of accomplishing these functions? It's the function, not the tire, which is important.

Let's consider oral care and teeth brushing. What is the ideal result? From whose perspective? Most of us might say, after some TRIZ prodding, something like, "the teeth take care of (Clean? Whiten?) themselves" as opposed to "the teeth have fewer cavities." But is that what the dentist would say? The percentage of dental income coming from traditional cavity treatment has declined dramatically due to fluoridated water supplies while income from procedures relating to appearance and prevention has increased dramatically.

Dentists are in the oral care, not the cavity filling, business today. If they had not made that transition, many would be out of business. How has all of this

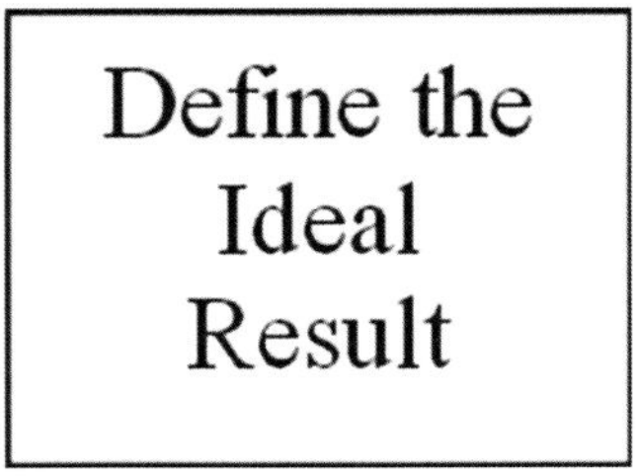

Fig. 4.5 The first step in
TRIZ thinking

> # Define the
> # Ideal
> # Result

affected the supplier of mercury amalgam or porcelain that is used to fill teeth? The manufacturer of great tasting, but high sugar level candy? Now we all might agree that the first definition of the Ideal Result is correct from a societal standpoint, but I use this example to illustrate the point.

Think about a hospital and its operation. If we confidentially surveyed nurses, patients, doctors, hospital administrators, and insurance company reps, would we get the same definition of the Ideal Result for the hospital? We certainly wouldn't. In this case there may not be such a clear cut answer to the question. TRIZ does not answer this question directly, but its lines of evolution (to be discussed later) may predict how the system will evolve, independent of individual opinions. That analysis may assist in guiding the discussions about the best interpretation of the Ideal Result.

In summary, the first step in this process is to clearly state the Ideal Result and view it from not just your perspective but also from that of your customer and your customer's customer. There are some people who just like to drive, but otherwise, what's the *function* of the car? To get us from place A to place B? Why? What *function* are we trying to accomplish when we do this? Meet with people? (Do GoToMeeting® or WebEx™ use cars or tires?) Shopping? Pick someone up? Inspect something? Go out to eat? I'll bet you are already thinking of other ways to do these without a car (and its associated tires). Now *both* Ford and Michelin need to be thinking about the Ideal Result. You saw how difficult it was to do this without some serious reorientation of our thinking (and maybe thinking like a kid for a while!).

We will gradually develop a graphical version of a simple TRIZ process for your use. We have now developed the first step in the process (this assumes that a readily usable answer was not found in a parallel universe search) (Fig. 4.5):

Do not underestimate the importance of this first step and doing it as an *independent* first step. This is similar in concept, but more extreme, than the independent idea (without criticism) generation part of creative problem solving or the "green hat" "what's good" step in DeBono's Six Hats™ process. The reason this must be an independent step is that as soon as we start to think about *how* to achieve this Ideal Result *before* we state it, we mentally compromise and what comes out of our mouth is something that is more like a 20 % improvement that we can easily envision achieving. The "how" compromises our vision of "what." It is absolutely essential to separate these two steps so that the Ideal Result can be clearly described, drawn, sketched, modeled, and elucidated. Without this first key step, the rest of the TRIZ process, algorithm, and tool kit will not accomplish what it is possible for them to achieve.

Exercises

1. Go into your kitchen or lunch room and take a look at a normal drinking glass. What's its *function*? What properties are optimized in the design of a glass? Color? Material? Breakability? Texture? What design variables are on your list if you were in the glass design business? Can we describe a more *ideal* glass from the perspective of several different types of people? What are you using the glass for? To drink something? Why? To swallow a pill? To refresh yourself? How could you accomplish these tasks or achieve these results without the glass? If you believe that this glass design may be more "ideal" in the sense of being able to hold it more securely without dropping it, how could this feature be made even more ideal? Here is a picture of one approach to this from govino®, designed by Joseph Perrulli and Boyd Willat (Fig. 4.6).
   Does this glass cost any more to make, once the original mold is made?
2. Pick up your coffee cup. What's its *function*? What would a more ideal coffee cup look like? Again, from whose perspective? What are the optimization variables? Depending upon how you answered the "function" question, how could that result be achieved without the coffee cup?
3. You are part of an engineering or product development team with an assignment to design a new system or product. What would be an Ideal Result description of your assignment? What kind of questions should you be asking about the assignment you've been given? How could the new product design *itself*? That's an interesting question isn't it? Are the quills up on the back of your neck? Why would you want to help someone do that? Since we *know* that's what is going to happen over time, we need to get on board the train! Have you seen all the TV news and weather channels that are now having people from around the country reporting news and photos? Ego (your name on TV) is a powerful motivator and is replacing a large number of field reporters. These "egos" also are free! The news reports *itself*. Where else do you see this process being used? Where else could it be used?
4. Consider your current product or service. How might your customer get the result that is achieved with your product or service without buying it? Suppose the use of your product or service was legally embargoed tomorrow morning— what would your customer do to meet the needs of *their* customers? Have you constantly updated your patent search related to their product line? What can you learn about the direction of their business development?
5. You are the CEO of a company and want to improve your capability to meet customers' unmet needs. What is the Ideal Result here? Is there more than one definition of this? How will you decide which you will use to take the next steps? Where will you find the people to give you the total and broad perspective that you will need?
6. What is the Ideal Result of your customer's customer? Do you know? Do you even know what your direct customer's customer does? What business they are in? What would happen if your customer's customer didn't need to buy their

Fig. 4.6 Govino® wine glass

product or service? How surprised would you be? Do you track the patent activity of *both* your customer and your customer's customer? Do you know how to?
7. If you are a manager with significant people responsibilities, including evaluations, forced rankings, salary decisions, etc., what process do you use and how much time does it take? Is it a lot of fun? If the Ideal Result is to fairly distribute a salary increase budget and have agreement between managers and those who report to them about how they're doing, how could this accomplish *itself?* Just pretend that all the time and paperwork involved in this process was not allowed to be used. How would you achieve the result?

Chapter 5
Identify and Use Resources

When we look at some significant inventions, we often see not breakthrough new technology but applications of very simple concepts in unique ways. These are the kind of inventions that, after you see them you say, "Why didn't I think of that?" Or "Why wasn't that invented decades ago?"

In the TRIZ world, what I am about to review is what we call the identification and use of resources and is the second step in the thinking process and algorithm that we are using. We have identified the Ideal Result and now we want to achieve it, but we do not want to spend a lot of money, use a large number of people resources, or a lot of equipment to achieve it. It's always good to use existing resources, isn't it? Why pay for something new if you've already got what you need? A great idea but the problem is frequently that we can't *see* all the resources we have in any given system. Many of us are using fabric bags to carry out our groceries instead of constantly using plastic or paper bags. We are asked to return hangers to the dry cleaner to save on disposal and new hanger costs. We recycle newsprint. We recycle aluminum containers to save the 80 %+ energy costs of smelting aluminum from ore. Recycling, however beneficial, is just the tip of the iceberg. Even at its best, recycling still uses energy to reprocess materials. In the case of a material like aluminum, this has a huge impact. In the case of plastics, these savings are not as great and the recycling process degrades the molecular weight of the polymer and its second and third time uses are restricted to less demanding applications. The real challenge is not to use anything in the first place. We see this at some large discount grocery chains where there are no bags provided and the customer carries everything out to their car in loose form.

In the TRIZ workshops that I am privileged to do for the American Society of Mechanical Engineers and the American Society of Chemical Engineers, we use a number of training examples to illustrate this point. A problem is presented and invariably the participants (many times engineers) want to *add* something to the system to fix the problem. After all, isn't that what engineers get paid to do—design things to add to systems that will solve some kind of problem created by some aspect of the current process or system? But then we've added something else that needs to be maintained and monitored. This may be a path to continuous

J. Hipple, *The Ideal Result: What It Is and How to Achieve It*,
DOI 10.1007/978-1-4614-3707-9_5, © Springer Science+Business Media New York 2012

incremental improvement, but it's not the path to breakthrough invention and innovation. *Adding* something to a process, whether it is mechanical or people based, always add complexity and complication. In the short run it may be an improvement, but in the long run it's a barrier to innovation.

Here's an example provided by Yuri Salamatov [1]:

> A robot was brought to a plant to operate a machine. After it was rigged up and switched on, the elderly worker who had operated the machine for years was amazed at seeing the nimble "iron man" performing all the necessary steps.
>
> A half an hour later, however, the robot came to a standstill, to the bewilderment of the service team of electronic engineers. What happened? As it turned out, some chips had fallen from the work piece into the moving elements of the machine. This situation where a human worker would simply flip the chips away with a broom and allow the machine to continue working brought the robot to a dead stop. The engineers cleaned the machine with a broom, switched on the robot… only to see the robot stop again after another 30 min. How could this problem be solved? Obviously, one cannot attach a human worker with a broom to the robot…

When I open a TRIZ workshop with this problem and ask the group to address the problem, here are the two major classes of ideas that are suggested. These are always independent of the experience or background of the individuals:

- *Add* a blower (or something else) to blow away the chips
- *Add* a shield (or something else) to prevent the chips from falling into the robot

Sometimes the suggestion that the chips be eliminated is suggested. That is indeed in the direction of an Ideal Result, but is out of bounds for this problem as presented because I specifically state that we have chips that need to be dealt with. The types of suggestions typically suggested would no doubt solve the problem, but they add cost and complexity as well as giving engineers something to design and control. Do you have any other suggestions?

How to deal with the chips?

1. _____
2. _____
3. _____
4. _____

Let's now take a look at this simple picture of a copper wire (Fig. 5.1).

Let's assume that we have a "problem" somewhere within or outside the area and volume of the copper wire, carrying normal current (let's say 15 A) and a "normal" voltage (120 V in the USA; 220 V in most of the rest of the world). If you want to pick some other level of these variables, feel free to do so. It doesn't matter what the "problem" is. We want to look for resources we have in this "system" prior to going outside it or spending money on something to solve the problem. Now I realize that the nature of the problem, and possibly the choice of voltage and current, will affect

Fig. 5.1 A copper wire

what resources you use and how you use them, but let's not get into that level of detail yet. Make a list of the resources you see:

1. _____   5. _____
2. _____   6. _____
3. _____   7. _____
4. _____   8. _____
5. _____   9. _____
6. _____   10. _____

Did you have trouble coming up with 10? Not unusual in my experience with workshop groups. Did you list "air" as a resource? (You should have!) If you did, why did you list air and not oxygen, nitrogen, argon, water vapor (humidity), etc.? Air is not air—it is made of multiple components. Is the air still? Does it have a velocity? Pressure? Viscosity? Heat capacity? Thermal conductivity? What's its temperature? The viscosity, heat capacity, density, thermal conductivity, and other physical all are functions of temperature and pressure of the air. What direction is the air moving across the wire? Is it changing? How often? It is unlikely that these physical properties of the air are uniform around the wire (since there may also be a thermal gradient around the wire if current is flowing through it), so these resources are *variable* as well. How fast is the air moving? What kinetic energy is there? What effect do these variables have on the thermal profile around the wire? There's not only a thermal resource but a *gradient.* Did you list the electric current? Most likely you did, but an electric current/field will also generate a thermal field (heat) and possibly a magnetic field. Did you list the wire itself? I am sure you did since it's the picture! But did you list copper? Probably. But copper (or any other material) is not 100 % pure. Did you list potential impurities? Copper oxide? Did you list surface properties? Surface roughness? Porosity? Is the porosity uniform? Impurities vs. depth of the wire? Length of the wire? As you can see, there are two to three times as many resources potentially available as you thought of initially. Did you list gravity? If not, why not? It's all around us and it's *free.* Atmospheric pressure is *not* zero. And it changes slightly all the time—we hear our weatherman talk about the barometric pressure all the time. Why? Because it's a sign of future weather conditions and the likelihood of sunshine, rain, or a hurricane. Again, how many of these resources and how you might use each of them is a function of the actual problem or issue, but in any case, seeing *all* the possible resources enhances our ability to solve a problem without spending money on *new* things. Most technical people are aware of indirect resources relating to their field of knowledge (such as the fact that an electric field will generate thermal and magnetic fields), but not necessarily all; and knowledge of these effects *outside* their field of knowledge is rare. One of the TRIZ readings in the references has a particularly good resource conversion table and is recommend to you (http://news.cnet.com/Gas-pipe-broadband/2100-1034_3-5945204.html). This resource conversion table lists the various scientific laws that relate to an effect or result that is desired or that can assist in figuring out how to convert one field to another.

Obviously, the fact that there is an effect that allows the conversion of one field or force to another does not guarantee that this will be an economical or desirable thing to do, but without this knowledge we are handicapped in our ability to think about and envision all the possible ways of converting one resource into another. This is why it is always important to bring into a problem-solving session individuals with totally different professional backgrounds and experiences or have access to a TRIZ resource which contains lists of fields and effects and their relationship to each other. Once when working with a company on an air traffic control display issue, I was amazed to discover that there was no knowledge of the literature or technical associations in areas such as chemical industry control displays, medical operating room displays, or video games. Tremendous resources for ideas that were not being considered.

Let's go back to our previous problem about the machining operation chips. What was on your list of resources? Probably the robot, the chips, and any other physical part of the system. Did you list gravity? If not, why not? Remember, it's free and it's always there and reliable. Now the original personal operator would care if turned upside down, but does a robot care? If we turn the robot upside down, the chips just fall away *all by themselves* using the *resources already present*.

In thinking about gravity as a resource, consider the catsup bottle. Why did it take decades for suppliers of catsup to realize that turning catsup bottles upside down, and using the force of gravity, would overcome all the consumer frustration of getting a thick viscous material to flow through a small opening, especially with little remaining catsup and the bottle having been stored upright in a cold environment, raising the viscosity of the catsup? And why did it take shampoo suppliers an additional 2 years to discover the same principle? Maybe because their product is not on the same aisle as catsup bottles or because they only go to meetings and conventions related to shampoos as opposed to "flowing of thick liquids."

This is the classical list of resources used in TRIZ analysis and problem solving, to which I have added one not normally found on a TRZ list:

1. Substances and materials
2. Time
3. Space
4. Fields and field/functional conversions
5. Information
6. People and their skills

We will review each in some detail.

(a) Substances/Materials

Substances are the materials within a system or the materials used to make something or the byproducts or wastes produced. We must include what we have considered "waste" in the past as many industrial examples of reusing and recycling what we used to throw away are all around us, from recycling plastics and aluminum to recycling of water in many industrial situations.

No material is pure. These impurities may have some utility. It is possible that minimal impurities may be the ideal and eliminating them, while possibly costing more, may greatly simplify a downstream process. Look at it both ways. Your material or product is being used by someone. Are you aware of the materials possibly of interest downstream? Is it possible that something you are removing (through the use of energy and labor) might combine with something downstream to save money and provide functionality for everyone?

Make a list of the material resources you didn't see before and force yourself to use them:

Material	How could you use?
1. _____	_____
2. _____	_____
3. _____	_____
4. _____	_____
5. _____	_____

Even when we think about normal store receipts, there is a large amount of information in addition to the amount spent. The nature of the product, when it was purchased, sales tax amount (if appropriate), store number, a toll free number for comments (possibly also providing a jackpot award for taking a survey), and in many cases in the restaurant business, a discount code for your next purchase.

Let's think about a normal drinking glass. If we needed it to hold water to dissolve something but didn't have water handy, what resource do we have? What about the saliva already in your mouth? Isn't that what we do with Listerine® breath strips?

We achieve the *ideal result* (the flavoring/breath scent) is achieved with *the resources we already have* (the saliva in our mouth—it's mostly *water*, isn't it?).

From my background in chemical engineering, looking at substance resources is similar to trying to maximize yield and conversion of a chemical reaction using the least expensive raw material available. From a managerial standpoint, we are beginning to see ads from a few manufacturers in the USA claiming they have no landfills. Everything that comes into the plant is incorporated into the car or sold as a byproduct. What would life be like if you had no waste disposal permits to worry about? How many fewer lawyers would be needed on your staff? How much less paper would be processed? What could you do with all the time and resources you used to spend on these tasks? Would you like to be in the waste hauling business around a plant like this? The fuel supplier to the hauling company? The tire supplier to the hauling company? The maintenance manager for the truck hauling company? The payroll supplier to its employees?

What kind of questions are you asking your purchasing people? Is price paid per pound of raw materials the most important question? What about price per pound of material used in a product that is sold?

(b) Time

We tend to think of time in the context of the process we are running, the task we are doing, or the service we are providing. As mentioned previously, there is time before and after the process we are running. The time a material spends in a truck or a tank car is a time resource. The time a material spends in a warehouse (including your supplier, you, and your customer) is a potential resource. The time spent driving to and from work, the delivery time for a product, and similar time segments are all time resources. It includes the time during the processing of a raw material we may be buying or the time during which a customer is using our product or process. The recognition that this is valuable time is reflected in the "just in time" process design approach that's currently in vogue.

When there is necessary inventory, shipping, or process time, we need to ask "what else could we use that time for?" When you actually see an assembly process where the parts are coming off a truck or another receiving line and *instantly* used in assembling or building something, you realize how much money has been saved in not building or renting space to store materials that will be used at a later time. In this case we *eliminated* time to make a system more ideal, but in the building products business, we use time during construction to allow concrete to "cure." (Curing concrete is a chemical reaction that is not very fast at room and ambient temperatures.)

Make a list of the time segments that are available to you and how you might use this resource:

Time segment type	How to use?
1. Time prior to your supplier receiving the material used to make your raw materials	_____
2. Time during your suppliers' processes	_____
3. Inventory time at your suppliers	_____
4. Transit time from your suppliers	_____
5. Time in your receiving area	_____
6. Time not used completely within your process	_____
7. Time after your process	_____
8. Time in your warehouse	_____
9. Transit time to your customers	_____
10. Time in your customers' process	_____

The "how to use" question may prompt additional questions such as "what do I need to make use of the time?" This then may prompt an additional, separate search and focus for resources necessary to make use of the time resource. How can the time be identified? Measured? You can see that many of the time elements listed above are negative in the sense that they are wasted time segments that not only consume time but also cost capital and maintenance expenditures. The use of "negative" resources is a subject we will discuss shortly.

Let's consider some other examples. The time before your plane actually takes off—how is it used? Load luggage? Search for drugs and explosives? Serve drinks to first class passengers? Check the airplane instrumentation? Make announcements to the passengers? De-ice the wings?

Do you or could you use "latent" chemistry in any of your products—a chemical system that is introduced into a product, but whose function is triggered later? An example of this is a dye that changes color when a food has been exposed to a higher than specified storage temperature. This example also illustrates the use of a negative resource (too high a temperature) to produce a positive result (change in dye color warning the customer that the food is no longer safe to eat).

To follow up on the "just in time" example mentioned previously, consider the Wal-Mart model of a customer sale instantly triggering a production and re-stock order to its supplier and shipping carrier. Time can be considered a "waste" or a "resource". Be careful not to judge too soon. If conducting an analysis to reduce time and inventory, be sure to ask if that time could be used in a positive way. In fact, use the proactive approach of forcing yourself to use the time segment in a positive way.

If you're a manager, how do you look at time? Is time with your employees a necessary evil or a resource to find out what's really going on without filtration from the managers in the middle? Is the time your salespeople spend with customers just to take orders and buy them lunch or dinner, or is it to find out what they are thinking about 5 years down the road? Do you spend time going to conventions and trade-shows relating to your current products and business or going to those that may cause you to think about alternative approaches and future challenges?

Time is a challenge to use and to define. In one sense, we need to ask how we can make better use of the time we have, and in another can we see time resources that we did not recognize or see before.

(c) Space

There is always space in and around a process or equipment. In today's climate, we are normally trying to eliminate wasted space and that is totally consistent with TRIZ thinking. Space is not just what we can see, but what we can't see. It's the voids and internal spaces within a solid. It's the surface, its geometry and texture, and the cracks. Not just on our product or within our system, but that in the raw materials we use and that of the system or product where our material is used. Think about food concentrates and juices that use the consumer's space and not that of the juice producer and grocery store. Think about the Matroiska dolls that you see that use the principle of "nesting" where space use is multiplied. As in our discussion about time, we frequently want to minimize space (as well as time as they both cost money in some form or another). As in the time case, if it is necessary to have a space, we need to ask, "How else could we use it?" There are voids in everything. Density is the physical property that we measure to determine this. Chemical elements such as mercury and bromine have extremely high densities compared to water's 1.0 (13 and 3, respectively). This allows large masses of these two materials to take up minimal volume. But consider a material such as a foamed plastic. These types of material have densities less than 5 % of water and are basically 95 % air or blowing agent. Recall our previous discussion about "negative" materials—materials that aren't there. That's what a void is—a lack of material. What else could the voids be used for? What could it contain that might be useful? To whom? We mentioned the fact that the wire we looked at before is generally considered to be a "solid", but as we mentioned before, *nothing* is totally free of voids. What could be put into the copper voids during its manufacture? What could it be used for?

Space is surface. There is nothing that is *perfectly* flat. Now that may be considered a defect in some situations (telescopic lenses, mirrors), but in most situations the lack of perfection presents an opportunity. Isn't that what sandpaper does? Surface roughness resulting from processing can also be used to increase adhesion or surface tension. It can be deliberately added, managed, and designed to create various types of sandpaper. It can "hold" materials. Surface pores may be a problem in skin care, but it also provides a place to "store" something. Space can also be characterized in length. We could just sit back and say there's no special resource here. Metal length is not constant. It changes with temperature (do you know your materials' coefficient of thermal expansion?) and with changes in mechanical force applied at either end. Thermocouple measuring devices use this property in a positive way. How could you make use of this longer or shorter (you could cool it, too!) length? Do you know the adhesion and surface tension properties of your materials? Why not? What if they changed significantly with temperature? (*They do!*) In your system, process, or product, what physical variable change would change the *amount* of a given space resource? What are all the variables that affect volume? How could they be used pro-actively rather than spending time and money trying to keep these properties constant? How could you use any normal deviations in a positive way instead of spending money "controlling" them?

Take a look at your cash register receipts. A long time ago, the only thing printed on them was the amount you paid. Then we listed each item by name. In some grocery store is now added is the amount saved by purchasing "specials" that day. On CVS drug store receipts (the longest I have ever seen!) are printed numerous coupon offers that are tied, in many cases, to what was purchased. This is a connection between space resources and information resources (to be discussed shortly). On the *back* of receipts from Walgreens you will also see coupons (did you ever think about the back of a receipt as a resource?). These can be linked to what was purchased or possibly to a competitive supplier of what was purchased.

Fig. 5.2 Space bag™

Probably anyone in the advertising business would say to us that *any* unprinted space is a resource. Voids (the *absence* of something) are resources. Using a competitor's sale as a positive resource is also an interesting thought. Anyone who has flown recently has seen the emergence of napkins and tray tables as sources of advertising space. Voids and hollow spaces can be used to store things. They can hide things. They can serve as buffers.

Most of us have seen the Space Bag™ product on TV and in the stores. This uses vacuum to withdraw air from the void spaces in clothing, greatly reducing the storage space required (Fig. 5.2).

Voids can be a positive! The *lack* of something can be a positive resource. Is this is a newly discovered phenomena (that there are voids in folded clothing)? And the system that is normally used to suck the air out of the voids in the clothing? A vacuum cleaner? Is this is a recently discovered invention? Are plastics that are excellent barriers to air transmission newly discovered? NO! This product, though "new," uses resources that have been around (but possibly not recognized) for decades. We now see this technology being marketed to reduce volume required for luggage used in airline travel with baggage fees being significantly increased. Charging for baggage is a relatively new phenomenon and this shows a good example of why it is good to go back frequently and ask about the Ideal Result and resource combinations that may produce new concept ideas. The variables that affect what we would consider an Ideal Result may have changed.

If you're a manager, what does the concept of "voids" suggest? Did someone just leave your organization? What *function* were they providing? Is it really necessary? Can it be combined with something else? Now there's a tendency in this type of situation to just add one person's job on top of another's and just overwork someone. That's not the kind of thinking I am suggesting. We have a tendency to habitually copy what has been done in the past. Our efforts in "lean manufacturing" are in the right direction as long as we think about *function*. We constantly see on the evening news examples of overlap and duplication in government programs.

This same kind of thing goes on in our organizations. Its *function* we need to think about!

What are the voids in your product line? How could they be filled? What *functions* (not products!) do your current customers need that you are not assisting them with? How could you supply this function? What kind of joint marketing opportunities could there be, allowing each company to make use of the others' resources?

Use this template to think about space resources.

Type of space	How can you use?	What affects? In what way?
1. Voids	_____	_____
	_____	_____
	_____	_____
2. Surfaces	_____	_____
	_____	_____
	_____	_____
3. Lengths	_____	_____
	_____	_____
	_____	_____

(d) Fields and Field Conversions

Fields are all around us, but sometimes we just don't see them. There's the heat or cold from thermal fields—a summer sunburn or the frigid winds of winter. There's that electrical current (field) that we can plug into. Is there a magnetic field around? Certainly could be. Have you thought about gravity? As stated before, it's all around us and it's free. It's amazing how often the clever use of gravity can substitute for expensive alternatives. There are also numerous simple scientific laws that govern materials and processes that we may not be aware of, so they don't even enter our brain for either problem-solving or new product concepts. One of my favorites is the Waissenberg effect. This is a law that says that viscoelastic materials such as bakery dough, as well as molten polymers such as polyacrylamide, do something very strange when a force is applied to them. When I ask workshop participants what happens in a rotating centrifugal device such as a washing machine, I hear the typical response of "the machine rotates, the clothes build up on the wall, the water goes out through the holes in the shell, etc." (And maybe a subtle question unspoken like "why are you asking me that stupid question?") Then I ask those same people to describe to me what they see when bakery dough is spun in the kitchen. The light bulb goes off and at least one person will say that the dough moves toward the center of the bowl and agglomerates around (clings to) the spindle (moving in the *opposite* direction of the applied force!). Then I ask what happens when the speed is increased. A few now remember that the dough actually moves *up* the spindle, *perpendicular* to the direction of the applied force! Not exactly an intuitive result. The exact same thing happens when a shaft is spun within a polyacrylamide polymer melt. This is not a mystery or accidental happening. There is a scientific law known at the Waissenberg effect which states that material with certain viscoelastic properties (like bakery dough or polyacrylamide) will move perpendicular to an applied force. The fact that this law is not as well known or famous as Ohm's law does not make it any less important in areas where it applies. If you're in new product development department at a major appliance manufacturer, you may never have even conceived of the possibility of clothes moving *up and out* of the washing machine. Here we see one of the basic limitations of psychologically based processes such as brainstorming. It is highly unlikely that a brainstorming group will suggest materials moving upward out of spinning machine because it is not possible to have all the knowledge of the inventors of the world in the room with you. The perceived "impossibility" filters the idea before it leaves our mouth. TRIZ is a process to fill in that void.

We start here by talking about the classical physical fields that most of us know— mechanical, thermal, acoustic, chemical, electronic, electromagnetic, and optical. There are, of course, subdivisions of these (for example, friction is a mechanical force also producing a thermal energy source), but for the purposes of this discussion, let's focus on these and consider each in turn. Notice that I said a thermal energy *source* (not problem). We will also discuss fields and field conversions.

1. Mechanical Fields

What comes to mind when you say this word? Bumping into something? A crash? Friction? Placement of an object by a machine? An assembly line and all of its components and parts? Something falling? Crushing? Compressing? All are correct, of course. The problem with mechanical fields is that many times we are trying to get rid of them. "Just in time" minimizes equipment movement. Friction is almost always considered a loss of valuable energy and in most cases that's true—we spend a lot of money cooling and trying to minimize it. Think about the millions of gallons of fluids used in the machining industry. One way to look at this, if you are a supplier of cooling fluids, is that this is a huge market opportunity. In addition, there are many possible opportunities for "optimization" (the enemy of innovation!) such as viscosity, surface tension, chemical stability, and corrosion. But another thought might be—what a waste of a resource! That heat is valuable—we spent money generating it and then we spend more money trying to control, manage, and get rid of it. The TRIZ approach would be to think about how to use it or preferably not generate it in the first place. (The Ideal Result thought would be how could be achieved the function generated by the device or machine producing the heat without its existence and thus without generating heat). The amount of mechanical force we have or use is frequently a function of geometry, height, and speed. Gravity can be considered as a mechanical field. It's all around us (and it's free) as mentioned in previous examples. In virtually all industries, mechanical control is being replaced by either electronic or electromagnetic fields, but that doesn't mean we should ignore it as a resource as it is readily available and usually relatively inexpensive.

Hydroelectric dams and pulley systems are probably our oldest examples of the use of gravity in a productive way.

As with all fields, a mechanical field or force has byproduct fields. A mechanical field will *always* generate heat—a thermal field. Are you trying to get rid of the heat or use it? Why did it take so many years for us to decide to use the byproduct heat from car braking to recharge the car's battery? A mechanical field might also create pressure. In fact that may be why we are using the field in the first place. If pressure is not a desired result, are you spending money trying to release and get rid of the pressure? Why not use it? Friction, involving both force and heat, is again, normally an undesired aspect of a mechanical field. We are normally trying to minimize it and there's nothing wrong with that, but could it be used in a positive way?

2. Thermal

When thinking about a thermal field, we are normally thinking in terms of heat input or removal (e.g., heating or refrigeration) or temperature (which is one way we measure of the amount of thermal energy). Sometimes we add heat deliberately to raise the temperature of a material to liquefy it or vaporize it. Sometimes we add heat to initiate a chemical reaction. We frequently have waste heat from manufacturing operations, chemical plants, conventional power plants, and nuclear plants. We spend a great deal of money in both cost and capital condensing or removing this heat and disposing of it to the environment. Our ability to reuse waste heat is determined to a great extent by basic laws of thermodynamics, but in many interesting industrial parks around the world, we find examples of companies collaborating in the use of waste energy resources.

A thermal field is also a potential indicator of the presence of a mechanical or chemical field. All chemical reactions, including corrosion, either generate or absorb heat, either one of which affects the surrounding thermal field and, indirectly, the temperature. Thermal fields are also used to indicate the recent presence of a "hot" piece of equipment such as an airplane or car. It can be used as evidence. Any temperature that is different than that of its surroundings has a differential thermal field and a signature that can be measured or used in a productive way. The value in using will be a function of the cost of energy at any given time and that again is another reason to continually review the Ideal Result and the resources thinking.

Thermal fields also generate byproduct fields including expansion and contraction, which automatically generate movement in materials.

3. Chemical

Chemical fields refer to the resource of heat and/or pressure change which occurs during a chemical reaction as well as the change in materials coming from a chemical reaction. Chemical reactions can be endothermic (requiring heat to sustain a reaction) or exothermic (giving off heat and continuing without additional energy input). In the first case, an endothermic reaction might be a place to use waste or byproduct energy from another part of a system. In the case of an exothermic reaction, we have the potential to use produced heat in another system. Coupling these kinds of chemistry provides one mechanism to utilize resources to their fullest. In a chemical complex, it is normal for great effort to be made to use the waste heat from an exothermic reaction to produce useful heat to run another reaction or keep a substance or area from freezing. Many reactions generate both pressure (a mechanical field) and heat, giving us two opportunities to reuse resources.

Chemical fields are what cause corrosion. Normally this is oxidation (rust) but it can also be fluorination, chlorination, or bromination. We usually try to eliminate corrosion. Corrosion typically expands the volume of a material. How could this be used in a positive way? Secure things in place? Indicate oxygen consumption? Fill a void? Create a force?

Do you like to clean your shower? Get down on your hands and knees and scrub the buildup of dirt and mold off the walls and floors? The Scrubbing Bubbles® shower cleaner from S.C. Johnson releases shower cleaning ingredients into the shower while you shower, eliminating the need for most "down on your knees" shower cleaning chores. You are running the shower anyway. The water resource you need is already there. And of course S.C. Johnson supplies the refills.

(e) Information

Information is a critical resource. In today's world, we almost have much information. We spend more time *filtering* information vs. collecting it. Information is not just what we think of classically data. Data such as temperature, speed, density, rate of loss or increase of a variable in processing are all "information" if we can relate this data to a desired or undesired result or action.

Two simple examples. The ability of Wal-Mart to maintain inventory and costs has been reported in many business publications. One of the keys to this success is use of information. When a sale is made, it immediately triggers purchasing and distribution vs. an employee taking inventory at the end of the day or some other data collection process that is not in real time. This is more of a normal practice for many businesses and retailers today, but recognition of this informational resource decades ago was a significant competitive advantage.

With access to the Internet and sensing technology, it is now becoming possible to remotely diagnose mechanical problems in appliances. In a recent example of combining a problem identifying "itself" and the use of information and technology resources, Sears® and its appliance subsidiary Kenmore® are pioneering the use of telephone-based diagnostics to allow technicians to analyze and narrow down an appliance problem and either eliminate or minimize the service time. See http:// wn.com/Kenmore_Appliances and watch a video of this technology. The capital investment in its appliance fleet is a major cost factor that can be minimized. The ability to generate and transmit this type of information over communication wires, as well as wirelessly, emphasizes the importance of continuously revisiting the list and capabilities of system resources.

Here's another interesting example of the identification and use of informational resources in problem solving. Imagine that you are the head of a police department in a major city. Part of your responsibilities and daily nightmares is dealing with traffic accidents and rushing police cruisers and possibly medical vehicles and equipment to the site of the accident. There are hundreds of potential accident sites. You know where many of them might be, but you can't afford to have resources standing by at every potential site. And of course the city continues to cut your resources every year, expecting you to do more with less. You're sitting in a conference room with your staff, trying to come up with new, inexpensive ways to monitor traffic accidents. Someone suggests more traffic helicopters and you quickly recognize the huge cost of this and dismiss the idea. You go through the list of TRIZ resources and pause at the "information" resource. What information would tell you that there might be a major accident somewhere? What do people do (now, not necessarily 20 years ago) when they are stuck in a traffic jam? *They make a cell phone call!* It will probably be to their spouse ("I'll be late—there's some kind of a problem on State St…"), their kids ("I'll be late to pick you up—I'm stuck in a jam and I don't know how long it'll be…"), or possibly a business colleague ("I am going to be late for our lunch meeting—will keep you posted…"). Unless you are in a car close to the accident, you have no idea why there is a delay. Though you may be curious about the cause, your immediate concern is to contact someone and update them. *That causes nearly everyone to make a cell phone call.* And what happens when someone makes a cell phone call? *An electromagnetic signal, transmitted through the air, is generated!* This is an example of field conversion that we discussed earlier. The mechanical act of texting or the sound wave of a voice is converted into an electromagnetic field that we can measure. It would be nearly impossible, in any practical or economic sense, to measure the mechanical or sound wave that was generated inside each car.

What if there was a way of analyzing and totalizing there signals (not their content, but their *amount*)? Wouldn't that tell you that something was going on? And you'd be able to send help before someone in direct sight, who might assume someone else had already called, might dial 911. And, of course, the people who can see the accident may or may not have cell phones and may be out of their cars trying to assist. The people not within view are going to call no matter what! A company in Atlanta, AirSage™, advertises itself as the largest collector and aggregator of cell phone signals in the USA. It has developed a service that recognizes this and provides this information to municipalities and emergency responders to provide an early alert to an emergency condition. Here is a quote from their web site (http://www.airsage.com/site/index.cfm?fuseaction=home.main&vsprache=EN).

"AirSage's Wireless Signal Extraction (WiSE™) technology aggregates, anonymizes, and analyzes signaling data from individual handsets using the cellular network, determines accurate location information and converts it into real-time anonymous location data. In essence, each individual handset becomes a mobile location sensor, allowing AirSage to identify how phones move over time."

This type of information, since people are on their cell phones for extended times, can also indirectly indicate the speed of traffic through the measurement of how fast the originating signal is moving! This provides yet another reinforcement of the need to go back and relook at the list of resources that are available to us. Twenty years ago there was insufficient use of cell phones to consider this as a usable resource.

Here's another interesting use of publicly available information that may not have occurred to you, but in principle would prove very valuable to speculators in the mergers and acquisitions area. When executives are negotiating deals, what do they do? Among other things, they use private or corporate jets to fly to have private meetings with their counterparts at the acquirer or about to be acquired company. When people fly, regardless of whether it is public or private, what are they *required* to do? *File a flight plan!* In a Wall Street Journal article (6/17/2011, pC1), the use of publicly available FAA flight plan information was used to predict a potential merger between Quest© and Century Link™. The suggestion is made that someone will make a business out of selling this type of *public* information. Incidentally Century Link™ did acquire Quest©!

How much information is generated by your process or your actions? What is done with it? What *could* be done with it? Are you measuring what you really need to know? What else could you do with all the information you are currently collecting? What information could be gathered from the fields being generated or used? If you're a member of any number of frequent flier, car rental, hotel, or travel sites, you know that your usage of these services is being tracked constantly. This is not only to deliver to your "rewards" but also to track your behavior and preferences so that special deals can be offered to you in a timely way. With the availability of multifunctional mobile devices, these types of things can be delivered instantly whenever they are needed (sound like an Ideal Result?).

There are many such examples of adding useful functionality in products, but let's also consider our business and marketing strategy and how this type of thinking can add value and make our primary function more ideal.

What activities are you and your team members involved in that could accomplish additional useful functionality? When calling on a customer, what else could we learn? Here are some questions to consider that will stretch the value of customer contacts beyond thanking them for past business, discussing future pricing, product shipping dates, etc.

1. What (not who) is a threat to your business? How can we help you meet that threat?
2. What kind of added utility or function could be added to your product or service? How could we help you?
3. What kind of things could happen (other than price changes) that would stop you from buying our product?

When there's a staff meeting, could we use it for other functions in addition to passing on information and reporting on standard activities? Here are some additional "added functionality" questions to use during your meetings

1. What is not on the agenda that we should talk about?
2. What topic is there that no one wants to talk about? Is afraid to talk about?
3. What could happen that would drop our sales 50 %? What would we do? What should we be doing now to prevent that?
4. What do we know about our customer's customer? Supplier's supplier?
5. What if our key raw material was suddenly embargoes tomorrow? How would we replace its *function*?

Note that many of these questions draw on the TRIZ basics we have already discussed. Make any meeting you have become an informational resources and not just an information exchange.

When thinking about information resources, use the following general template:

Functions being performed	Information being generated
_____	_____
_____	_____
_____	_____
_____	_____

Many resources in the previously discussed categories can also be considered "informational" resources if we broaden our thinking sufficiently. For example, friction and wear that produce elevated temperatures and possibly noise (a field) can tell us that maintenance is required for a piece of equipment. The mechanical field action of someone writing a comment down on an evaluation form is a source of informational feedback that will tell us how to improve a workshop. A color change (chemical field) gives us information as to whether a frozen food product has thawed in storage or processing. The movement of a compass needle (a magnetic field response) gives us the information we need to know what direction we are heading.

What are your examples of fields that are generating information?

Field	Information generated	Useful for?
1. _____	_____	_____
2. _____	_____	_____
3. _____	_____	_____
4. _____	_____	_____
5. _____	_____	_____

Total knowledge of field conversions (all the different byproduct fields generated by a given field or determining a physical law that defines how one field or force converts to another) is impossible for any one person to have. There are many resources, in print, the web, and in various TRIZ software products, that provide assistance. See the appendix for references and resources.

(f) People and Their Skills

This section is a direct result of having been challenged, in one of my public workshops, on the fact that the traditional list of resources in TRIZ was incomplete. It didn't mention people. This gave me real pause as I recognized this as a major deficiency in TRIZ analysis of resources. If we just think about numbers of people, one obvious use of this category is to make sure that surplus people resources can be used somewhere else in a process or system. For example, this might point toward crossfunctional training to be able to utilize our human resources to the greatest extent under any business conditions. But let's go deeper. I now ask everyone in my workshop this question. What skill do you think I (meaning Jack Hipple) have that is not obvious to you? Now those of you who know me realize that the answer to this question has nothing to do with any physical attribute. Occasionally someone will guess the answer and that is that I am an amateur musician (but not that I am a French horn player). While at the Carnegie Institute of Technology (now Carnegie Mellon University), where I received my degree and education in chemical engineering, I was able to continue my high school music education and play in the famous Kiltie Band. I now play in a few adult community bands whose primary goal is to make residents of retirement and nursing homes feel better. My skills are not sufficient to make a living this way, but it *is* an interest and unusual skill that I have.

Another interesting example of this was shown in a Public Broadcasting special a while back. In order to assist in breaking German codes during World War II, British intelligence hired excellent crossword puzzle workers who could see patterns in words that no one else noticed.

Here's the question you need to ask every one of your employees: "What skill or interest do you have that we are not using in a positive way?" and "In what way could we use your talents more effectively?" And possibly a third one, "What would you really enjoy doing that you aren't doing now?" This will give you some new insights, provide for some interesting conversation, and give you some new ideas for special assignments, career growth, and training resources. In times of high unemployment and pressures to conform to keep our jobs, people are hesitant to speak out and express themselves. Encourage them to talk about themselves and their skills, especially those you are totally unaware of or are not using to *your* advantage and *theirs.*

Our emphasis on thirty-second elevator speeches, not reading anything over two pages, and pitching only what is desired to be heard pushes us toward an environment of *less* information about people, their skills, their interests, and their talents. There may be times when this is appropriate, but we need to be careful.

(g) "Negative" Resources

Before looking at a final example problem that pulls together the two concepts of Ideal Result and Resources, let's consider the general concept of a "negative" resource. This can be any one of the previously mentioned types of resources. Take for example the heat generated by a chemical reaction, the heat generated by braking a car, or the heat generated by a computer data center facility. In all these cases, we use some kind of a cooling system to remove the heat at some expense. If we can find a product or another system that can use that heat, then we don't have a problem—we have an opportunity to use a negative resource in a positive way. As mentioned previously, this is the concept used by Toyota in using braking heat to recharge the battery in their hybrid vehicle. We can conceive of industrial complexes that are grouped around byproduct energy production and the need for energy. One plant's "negative" resource is a positive resource for someone else.

Keeping that thought in mind, let's take a look at combining the concepts of Ideal Result and resource identification and use by considering the following problem:

> It is necessary to separate some defective pharmaceutical pills from perfect pills. The defect is a dent on the circumference of the pill. Now we could, and should, make sure that there is nothing we can do upstream in the process to eliminate this defect, but let's assume for the moment that this is not possible or that the cost is prohibitive. What might we do?

How might we approach this problem? It's easy and dead wrong to jump into the solution space or start brainstorming without going through the TRIZ algorithm, step by step. So what is the Ideal Result?

One answer is not to make the defective pills in the first place, so a relook at the process is certainly warranted. But if we are forced to deal with the problem as one that exists, then how do separate the bad pills from the good pills? When this problem is analyzed by experienced engineers, ideas such as weighing the pills (to detect the difference in pill weight reflecting the fact that some of the pill material is missing) or some kind of flotation system (to take advantage of how the non-smooth pills might behave differently in a flowing air stream) are frequently suggested. These might work as well as the original suggestion of a laser video inspection system to spot the differences in circumference uniformity. But they are all expensive and limit the economic viability of the process change and eliminating the human labor used. How can the pills "separate *themselves*" using the resources *that already exist?* What resources do we have that are readily available and at minimal cost? Do you remember gravity? What would happen if we turned the pills "on end" (just change the orientation of the discharge of the pills)? In practice, this is exactly what was done. At the same time, the gap between the 45° belt and the horizontal belt was changed from virtually zero to a few inches. The pills which have perfect geometric circumference roll down the belt and have sufficient momentum to "jump the gap" on to the final belt, while the pills having circumferential defects fall over and, upon reaching the end of the 45° belt, simply drop into the trash can all by themselves. *The bad pills separate themselves using the resources that already exist.* Recall the previous discussion about "negative" resources. The defect in the pill that we used is actually a negative aspect of the system that we used in a positive way. This is very close to an *Ideal Result*. In a workshop I ran a while back, one of the participants actually took out an aspirin tablet he carried with him and modeled this solution by rolling the pill down the slope of his three ring workshop notes binder!

In TRIZ problem solving, it is *always* worthwhile to make a separate list of the negative things about a system and force ourselves to think about how to use these things in a positive way.

Let's think about "negative" resources and how they are used. Have you observed wear indicators on tires? The wear level displays *itself* via unique design of the tread. In the Toyota Prius™ automobile, the heat generated by braking, instead of being considered as a thermal waste that must be "handled" or removed, is used to recharge the battery. Do you use a stain prior to brushing your teeth that shows vividly where proper brushing has not occurred? The *negative* is a *positive* training data point. The mistakes noted on a test show where we need to improve. The number of failed logins indicates possible difficulties for consumers in using web sites or possibly incidents of security breach attempts on the Internet.

A major problem in drug treatment therapy is the too rapid release of active ingredients after swallowing a pill. Accuform™ is a technology developed by a specialty pharmaceutical company, Depomed®. The technology involves coating a drug ingredient with a proprietary mixture that swells in the presence of gastric acid in the stomach. This increases the size of the pill and prevents it from passing immediately into the gastrointestinal tract, allowing a much slower, controlled rate of drug release. Using a negative as a positive! We could conceive of this approach being useful in the development of a drug delivery system for diabetes if we could coat insulin with a material that would swell or convert to a protective layer using the gastric acid in the stomach. The inability of insulin to survive chemical attack by insulin is the primary reason that diabetic patients must use insulin injections that allow the insulin to go directly into the blood stream.

Here's another exercise for you. Make a list of all the *negative* material resources (such as impurities in a material), negative information/feedback, and negative fields and force yourself to use them in a positive way. As an example, consider how we are now using byproduct methane from landfills to generate steam and electricity, while at the same time, eliminating a potent greenhouse gas.

Make this list for your system:

"Negative" aspects	Use this for:
1. _____	_____
2. _____	_____
3. _____	_____
4. _____	_____
5. _____	_____

   Resources can change and it's good to revisit our list and vision of resources over
time. Consider Amazon.com© and how it has one major warehouse in Nevada,
replacing potentially thousands of individual bookstores all around the country.
I recognize that there are passions around the topic of small family businesses vs.
large businesses. TRIZ doesn't answer these questions that have a moral or emo-
tional component. It does clearly say that systems will integrate into a super-system,
just like the bucket integrated into the paint stick and the tire integrated into the
wheel. Politics and emotion may affect the rate at which this happens but it is an
irreversible trend. What resource did Amazon.com see and take advantage of that
others did not see or took much longer to see and utilize? *The Internet.* One huge
warehouse and the information resource of the Internet replace thousands of small
bookstores. Who else benefits from this move? The US postal service as well as
companies such as UPS© and FedEx©. Did they wait for Amazon.com to call or
did they put on their TRIZ hats, recognize what was going to happen, and approach
the Amazon.com founders?
   The previously cited example from AirSage also illustrated the use of commercial
satellites and the Internet, resources that simply did not exist or were not readily
available 20 years ago. In addition to updating our resource list occasionally, it's
also a good exercise to keep handy a list of resources that we "need" to achieve an
ideal result and can't find or identify and keep thinking about how we could find
them or convert other things to them.
   The need for new rights of way for pipelines and other transportation mechanisms is a constant need. This is true for not only oil and gas pipelines but also cable
and fiber optic lines. What if we used the existing infrastructure of our interstate
natural gas distribution network that is already in place? (http://news.cnet.com/
Gas-pipe-broadband/2100-1034_3-5945204.html).

Exercises

1. Look around your office or the room where you are sitting right now. Think
   back to the wire example and make a list of all the resources you see in the room
   that might be used as resources for any problem you can think of. Think about
   resource conversion that we discussed (mechanical to thermal, electrical to
   thermal or magnetic). What other resources could you create? Someone just
   walked into your office by opening a door. Would that affect your answer?
   What *new* resources were created? Someone just turned off the lights. What
   new resources do you have?
2. Another *person* just walked into the room. What's changed? What new resources
   do you now have? (In addition to the added body!)
3. Your new product development team just lost a key member who was hired by a
   competitor. What *new* resources do you now have to meet your team objectives?
4. A bottle of a chemical is sitting on a laboratory bench. What resources does it
   possess? How would this be affected by the type of chemical?

Fig. 5.3 Combining ideal result with resources

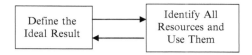

5. You are sitting on your couch at home. You've been asked to assist in sweeping the rug in front of you. To avoid having to do this the next time, make a list of the resources you can employ to assist in the rug cleaning "itself."
6. On your last customer or supplier visit, what did you see or hear that you ignored because you were overly focused on what the current business and product concerns are?
7. What skill do *you* have that is not being utilized or appreciated? Are you going to wait until retirement to use it? Have you said anything to anyone? Why not? Is the decision to wait to use it the best one?
8. What is the most irritating thing that you can think of right now? How will use it in a positive way? What waste material or energy are you spending money to dispose of? Who could use it?
9. Make a list of all the fields in your current process. What are the byproduct fields/forces this primary field is generating? Are you using them? What byproduct fields are generated you are unaware of?
10. How could you use gravity in your process?

We now have the second box in the process (Fig. 5.3).

Note that I have diagrammed this as a recycle loop, indicating that the discovery of additional resources that we were not aware of at the start may change our definition of the Ideal Result in that we will envision opportunities that we did not think of originally.

Reference

1. Salamatov Y (1999) TRIZ: the right solution at the right time. Insytec, The Netherlands, p 3

Chapter 6
Whose Ideal Result and Whose Resources?

We need to take a little side journey at this point to discuss in more detail something we alluded to earlier in our discussions about the Ideal Result. In that discussion we raised the issue that might arise in a medical care situation, that all the stakeholders may not have the same view of the Ideal Result. It is very rare that in the real world everyone will agree on the Ideal Result's definition. Bosses and subordinates will not. Parents and teenagers will not. Functional representatives on a team (manufacturing, research, marketing, etc.) will not. Corn growers and corn consumers will not. Oil producers and oil consumers will not. Using TRIZ requires that we have a definition of the Ideal Result. In a case where there may be differing views of this, it may be worthwhile to use the tools directed at the various end results and see where we end up and see what commonality there might be. Unless we are talking about a situation where someone's view of an Ideal Result violates some law of physics, this discussion is significant and important to have as the discussion itself will generate new concepts of approach.

The same can be said for the identification and use of resources. Not every stakeholder in a problem situation will recognize or value resources in the same way. For example, complaints filed against a health insurance company could be viewed negatively by a rating agency or positively as a source of information about how to improve service. I have taken cars in for service at a number of car dealerships who make a practice of telling you that, if you are satisfied with the service they provided, you should fill out the survey form Emailed to you by the parent automotive supplier as "outstanding" or whatever the best rating category is. Why do they do this? Well, in many cases, any response short of "outstanding" generates a call from headquarters. Now when I take a car in for service, I *expect* everything to be done as requested. This is *normal* not *outstanding*. Outstanding might mean that they washed my car without charging me and fixing some other minor thing while the car was there without being asked or my being charged. By telling the customer ahead of time what the "answer" is, true feedback is eliminated.

When we come to the later discussions about contradictions that stand in the way of achieving the Ideal Result or in the use of resources, we will also encounter this challenge as everyone's view of contradictions will not be the same.

J. Hipple, *The Ideal Result: What It Is and How to Achieve It*,
DOI 10.1007/978-1-4614-3707-9_6, © Springer Science+Business Media New York 2012

Let's go back to health care. I suspect that we will never get complete agreement on the Ideal Result from all the parties involved, especially as the government becomes more involved in health care policy decisions and reimbursements. Politicians desiring re-election, and having no qualms about postponing debt, will display no hesitation to raise benefits and reduce premiums. Hospitals will probably always need patient advocates with differing views of what the appropriate care for how long is constantly debated. Medical malpractice attorneys will never see things the same way as the surgeon in a delicate, risky surgical procedure. TRIZ will not give answers to these issues directly, though some of its forecasting lines of evolution, which we will discuss later, can provide keys to seeing whose definition will eventually prevail.

The utilization of resources which we just discussed can have the same challenges between stakeholders in a problem. In this area, it is more often an issue of being able to *see* things differently as opposed to having views about resources being right or wrong. In a few cases, there may be moral or ethical values attached to the use of certain kinds of resources and TRIZ provides no resolution to these differences in opinion.

Consider the area of oral care which we mentioned earlier. Let's make a table of The Ideal Result and resources readily seen or recognized by the various stakeholders involved in oral care, defined fairly broadly (Table 6.1).

Let's be honest here. Is it the Ideal Result for the dentist to have zero cavities? Or is it to maximize income? Does the patient with no cavities also want to maximize the income of the dentist? What if the patient's Ideal Result (let's assume it is perfectly white teeth at all times without ever going to the dentist's office and never receiving a bill from the dentist or a dental insurance company) resulted in *no* income to the dentist at all? What if there was a self-diagnostic tool, usable in the home, and "Do it yourself" cavity and enamel repair kit available at the local drug store? As fluoridated water has eliminated a large percentage of traditional cavities, the dental profession has moved to more cosmetic services such as teeth whitening and replacement and straightening of teeth.

It is also interesting how different stakeholders in the health care business see resources in different ways. No one wants to yield their special expertise to others. Dentists will argue today about how signs within the mouth, such as saliva, gum condition, and tongue characteristics, can be predictive about other health conditions in the body. Medical doctors are reluctant to admit this as it may take away some of the need for their special expertise and allow new OTC ("over the counter") diagnostic tools not requiring prescriptions written by a doctor. Dentists see the mouth as a resource; most traditional medical doctors see it as a delivery system for drugs or scopes.

Nurses are considered valuable resources by many doctors, while some others view nurses as expensive technicians. TRIZ is not going to resolve these conflicting views, but can provide a predictive view of whose views might ultimately prevail

Table 6.1 Differing views of resources

Stakeholder	Ideal result	Resources
Dentist		
Patient		
Dental technician		

(more later when we discuss lines of evolution). The point here is just to sensitize you to the different perspectives on these two subjects as you try to apply TRIZ in the real world. If nothing else, this discussion with your fellow problem solvers in this area will make everyone aware of the motives and perspectives of everyone involved.

Let's take a broader view of the different views of resources that might be present in the health care area. If you're the hospital administrator, here's what it might be:

1. Space: the hospital facilities, the beds, equipment
2. Time: time the patient stays in the hospital (which *part* of the hospital?), time between billing and payment, time prior to patient arrival
3. Substances and materials: Purchased supplies (linens, gowns), bank loans/money
4. Fields and field conversions: Cleaning chemicals, prescriptions, gas cylinders, antibiotics, patient movement, alarms, electrical and magnetic fields generated by equipment, pacemakers, etc.
5. Information: Previous health history, patient data that is monitored, insurance reimbursement information, allowable stay parameters, patient knowledge of conditions
6. People and skills: Surgical and medical knowledge of staff, nurses' training and experience

What if you're the patient? Is the list the same? Even if it were, would it be in the same priority?

1. Space: the room, the bed space, the space around the room, the view outside the window in the room, the pockets in the hospital gown, the bedside table space, the feeding table space, and layout
2. Time: 24 h in the day or the time the patient is in the hospital (not just the hospital room!), time waiting for food, the "on call" button to be answered
3. Substances and materials: Reading materials, personal effects, clothing, money, insurance policy, body composition, and feelings/pain
4. Fields/field conversions: Illnesses causing fevers or chills, walking or movement causing pain, pressing call button to send an electrical signal
5. Information: status of treatment, pain level, time of day, number of nurse visits, performance level of hospital in a particular specialty
6. People and skills: knowledge of medical history, technical knowledge of disease or injury, communication capabilities

What's the view of the doctor?

1. Space: Adequate space and privacy for patient visitation, adequate operating room space, space for sleeping (if you're a resident!), space for writing reports and notes
2. Time: Adequate preparation time for surgery, to review patient information and history of supporting staff, time in transit to/from the hospital or office
3. Substances and materials: All surgical and medical supplies needed, emergency surgery supplies, interfering medications
4. Field/field conversions: Movements, conditions that cause pain, interfering instrument signals, magnetic fields, cell phones, implants

5. Information: Patient history, medication history, relative medical history, past surgery information, historical knowledge of illness
6. People and skills: Experience of surgical nurses, personal success with a particular surgical procedure or illness

And what about the insurance carrier?

1. Space: office, size of hospital, size of rooms, real estate for claims processing
2. Time: billing turnaround time, time value of borrowed funds, status of premium payments, time for diagnosis and approval of surgery or special care
3. Substances and materials: paper and electronic information, money
4. Fields/field conversions: bank transaction time, "float," credit conversion into assets, hospital conditions possibly creating liability situations
5. Information: patient's health history, hospital cost records, doctor cost records, liability history, payment histories
6. People and skills: billing staff history and understanding of hospital stays (has anyone summarizing a bill ever had a hospital stay?), medical knowledge of staff

Now, you may not like my list so go ahead and make your own. The important point here is to see what these various groups (including ones not mentioned like families, hospital goods suppliers, ambulance providers, etc.) might have in common. I think we might agree that accurate patient health history (an information resource) and accurate information on the condition of the patient who is hospitalized would be in common. Without saying that the other issues are not worth looking at, we could safely say that anything we do to make health information and history more "ideal" is going to happen, independent of who is defining the Ideal Result. This fact might provide a nucleus of common ground for an Ideal Result.

Exercises

1. Your teenager has just asked you for permission to go "steady" with someone. What are some of the differences in the Ideal Result here as viewed by you, your teenager, their new steady friend, and the parents of the new "steady"?
2. Your company has just spent a small fortune on a new manufacturing line using some new, specialized, automation machinery that, if run properly and efficiently, will eliminate half of the current workforce. This new equipment is to be run by the same group of technicians that ran the equipment to be replaced. What is the Ideal Result from both the managerial and operator perspective? Could you design a process that would meet both of these definitions equally well?
3. You are the chief engineer with a major copier machine manufacturer. It is technically possible to add more features to your current machine. However, you know, from consumer feedback, that many users have no idea how to use the many features that your current machine already has. How do you, the customer's office manager, the office assistants, and the casual user view the Ideal Result?

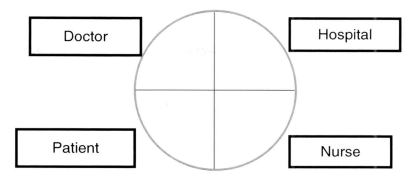

Fig. 6.1 Satisfaction for ideal result

4. One of your customers' purchasing agents has just informed you that they want a 10 % price reduction in a contracted purchase from you next year. It's clear what their Ideal Result is—they told you! Yours was probably to raise the price, right? What kind of approaches can accomplish both Ideal Results?

5. You are the operator of a profitable landfill. Your long-term profits are projected to decrease as its space fills up and costs of capping and closing increase. What resources do you have that might lengthen its life and prolong, and possibly increase, its revenue? How does the surrounding city or county view this situation? What could they do with totally sealed and contained landfill? Do they view the resource the same way you do? Is there a way to keep everyone happy?

6. You are the owner of a major airline with unpredictable fuel costs, varying passenger loads, and changing FAA regulations. There are new jet engines coming on the market which will reduce fuel costs by 20 %. Your labor unions have granted you "givebacks" in the last contract that have allowed you to remain profitable. In this situation, what is the Ideal Result as viewed by the airline management? The new jet engine supplier? The maintenance employees? The maintenance employees' union? The flight attendants? The flight attendants union? The pilots? Pilots' union? What suggestions do you have that would maximize the percentage achievement of each Ideal Result?

7. You have been asked to come into a company and run some training sessions on a new breakthrough problem-solving process (TRIZ?). You know, from previous contacts within the company, that there are already embedded processes that are popular and that have produced positive results. There are at least three different definitions of the Ideal Result here. What are they? How many more might there be and who owns them? How can you design the training to maximize the degree to which all of them are achieved?

8. You are the human resources manager of a Fortune 500 company which is instituting a new performance appraisal system, reducing the number of rating classifications and varying the time between raises. Before you implement this program, who are all the different people and groups affected and what is their view of the "ideal" pay for performance system? How do they see the various issues?

In thinking about *whose* ideal result we are talking about, and how well we have done, use the following plot of % Ideal Result versus stakeholder. The more of each segment of the circle we can fill in, the better we have done (Fig. 6.1).

Our ability to maximize (*not optimize!*) the area filled in is a measure of how well we have achieved the various stakeholders' Ideal Result. There are going to be situations where the Ideal Result will be defined by someone in such a way and under a given situation that simply will not satisfy everyone. The tools of TRIZ can still be used in this situation. Hopefully, the mandated situation will be reviewed at a future date to make sure that other viewpoints can be considered and analyzed.

Chapter 7
Adding Useful Complexity: One Approach to the Ideal Result

In the next two chapters we are going to discuss two TRIZ tools that, in a general sense, are contradictory to each other. They are both part of a long-term TRIZ analytical principle that says that products and systems oscillate between simplicity and complexity. We start with a simple, possibly not particularly special or unique, product or system. We discover that adding parts, options, appendages, and choices (technical or nontechnical), adds complexity. In many cases, we find that these added features are ones that consumers need (or think they need) and are willing to pay for. Then, at some point, we have a product, system, or service that is fancy, useful, complicated, and possibly relatively expensive. We may have maximized profits in the short term but we are on dangerous ground as we have created a complex, but useful, system which is begging to be replaced (Similar to a very sturdy, corrosion resistant paint tray just prior to the introduction of the Pivoting RapidRoller™). Now it's time to look at simplifying.

This overall curve looks like the one shown in Fig. 7.1.

Moving along either side of this curve can make a product or system more ideal. It's a matter of perspective, cost, competitive options, and the super system in which the product or system is used. We'll look at both approaches and that's what you should do in your own work. In this chapter, we'll focus on the left side of the cure—*adding useful complexity.*

The definition here is the addition of "something" to a system that improves its ability to produce a more Ideal Result (from someone's perspective)—its ability to perform something useful or make a product or system (or a person?) able to perform additional functions. The complexity must not make the system or device so complex to use that the usefulness that is added becomes itself a barrier to its use. It's best if what is added to something replaces something else (a product or service). This is the "trimming" principle we have already mentioned and will discuss more completely in the next chapter.

We've been a little harsh on "adding" things to systems, but we do need to consider this approach with some seriousness. Adding something useful to a system *can* make it more ideal, *especially* if it replaces something more complex or replaces another system altogether, making the overall super-system easier to use

J. Hipple, *The Ideal Result: What It Is and How to Achieve It,*
DOI 10.1007/978-1-4614-3707-9_7, © Springer Science+Business Media New York 2012

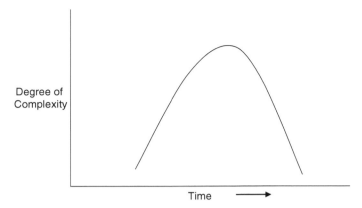

Fig. 7.1 Complexity versus time

or less complex. In many cases, what we see is the *combination* of products or systems that were previously separate. The shower cleaning and floor cleaning examples illustrated this. Consider the oral care product from P&G, Glide™. Previously we had these functions (gum stimulant, flossing material, etc.) in separate products and slowly they have now been combined into one product which probably uses, overall, less material and plastic. But that plastic or polymer is more sophisticated and functional. If you were the plastic or polymer supplier to P&G, what kind of questions were you asking your customer? How much more could I charge you before losing your business? Or were you thinking about the fact that people used two *different* products to accomplish these goals? How could you change the functionality of the material you sell, patent it, and then license the technology to P&G?

If two utensils can be combined into one, then less shelf space is required. Also, travelers just need to carry one utensil and use less space in a carry-on bag. This factor has become increasingly important as luggage fees have appeared in the travel industry. If one of them was a metal pick, that traveler no longer has to worry about getting their tool through airport security screening. Although the industry is totally different, the concept and pattern is no different than what has happened to office machine integration (copiers, fax machines, scanners in one device or system) as well as self-darkening glasses versus having two separate pairs of glasses or sunglass clip-ons.

We need all these functions and keeping track of all these simple things is not easy. Running out to the hotel, gift shop, or the local drug store on a trip can be a real inconvenience.

A Swiss Army knife combines many different types of knives and mechanical instruments and now recently has added a flash memory!

Up to a point, there could be additional items added to this device. What might they be? Remember, we're adding *useful* complexity.

What *else* could we add to a Swiss Army knife that would be useful?

1. _____
2. _____
3. _____
4. _____
5. _____

The "copy" machine that sits on your desk or down the hall from your office can not only copy but also scan, fax, and Email, replacing two or more separate machines. Office desk surface has been saved and can now be used for other purposes. As long as this added useful complexity is easy to use, does not add tremendously in cost, and uses less total space, this more complex device will provide "value" and will sell. You might ask yourself what other function could be added to this machine and then ask how this could be added without decreasing ease of use.

What additional *useful functionality* could be added to the existing office machine that you use?

1. _____
2. _____
3. _____
4. _____
5. _____

In each of the above lists, make clear what is being "replaced," so that we have a sense that we are moving in the direction of making the system more *ideal*.

In the food industry, we see the combining of products in a way that adds useful complexity to food products. Artificial sweeter manufacturers have begun to add vitamin supplements and fiber to their products, eliminating the need to eat two separate products to achieve the goal. Packaging is reduced as well.

Now let's take a walk down the hall to your "big job" copier machine and try to make one black and white copy. Can you figure out how to do this? Do you understand the control panel? Do you know what to do? How complex has this panel become? Is it useful? To whom? The occasional user or just the experienced office assistant or copier repair technician? Did you just try to find an office assistant to help you? Why didn't you have this problem with the smaller machine in your office? We'll come back to this problem later when we discuss TRIZ separation principles and the use of the other side of this complexity/simplification curve.

Grab your toothbrush and let's go back to the oral care area. Are you old enough to remember plain toothbrushes? All the bristles were the same length; there was no choice in handle length, bristle type, or any other physical aspect. The toothbrush head now has bristles of varying length and hardness. A manual toothbrush frequently has a flexible handle. Electronic toothbrushes, costing orders of magnitude more than simple manual toothbrushes have introduced high speed vibration to massage gums. The alternative to this is a more frequent visit to the hygienist or a *separate* tool and a *separate* time-consuming step. We have added useful complexity but at significant cost. However, this cost is perceived to be far less than the alternative of gum surgery, cavity fillings, or separate added tools and time. This is a case that must be watched as the cost difference is significant and the use of TRIZ tools such as trimming (mentioned previously and to be discussed next) and contradiction resolution (to be discussed later) might come into play in a replacement product or system. Many of the newer tooth-brushes have battery power embedded in their handles. This is certainly adding complexity. Is it useful? Most of us would say yes. It makes sure the brushing strokes are uniform and not dependent upon our skill at moving our hands up and down or our ability to maintain the frequency of movement. It can also be considered to be a low cost alternative to the rechargeable brushes. Some toothbrushes now have embedded color indicators to tell us when bristles are worn and a new toothbrush is needed. More useful complexity and again at an added cost. But, again, we perceive this cost to be less than the consequences of using a worn out toothbrush too long. In the oral care area, we have discovered so many long-term consequences of *not* brushing properly (correct way, correct time length, gums as well as teeth) that we are still adding useful complexity and paying for it. The oral care area as it has evolved over time illustrates many different TRIZ principles.

Of course, we can't leave this topic without discussing cell phones (or PDA's or whatever you call your communication device today). This device and what it does is another great illustration of the use of TRIZ principles. Let's just consider the phone and its functions. It used to just make phone calls. It made phone calling "wireless" and we could walk around the house, ride in a car, etc. without worrying about whether the call would follow us, etc. but at a significantly greater cost. This would appear to be added useful complexity. Your "phone" bill has increased substantially, but most of us have decided that this extra complexity is useful and worth the extra price. With the addition of satellite technology, the cell phone can do Email as well, saving the lost time in retrieving Email from your office or home computer. The "phone" can also now take pictures, in some cases eliminating the need for a separate camera. You can even save these pictures directly to your office computer without having to visit your local drug store. Thinking back to the last chapter, is this the Ideal Result for you? For the drug store? These pictures can be transmitted, via your Email, anywhere in the world. Many drug stores have enabled us to send the photos from a phone over the web and then pick them later. If you are an insurance claims investigator, this added complexity is truly useful and a real cost saver. If we step back a second, what has happened to your "phone" bill. If it's anything like mine, it's an order of magnitude more than the $20 from several decades ago. The added useful complexity has replaced the need for many other devices. They have been *integrated* into one product and system. In this case, the added useful complexity makes the product more ideal. Right now, this useful complexity (despite its cost) is still in demand in the market place. We have decided, subconsciously possibly, that the extra $180 is cheaper than whatever we used to do or whatever time we needed to accomplish all these functions.

In one of the most recent additions of useful complexity to this device, software products have been developed which allow a user to instantly compare prices by scanning bar codes in a store and instantly show comparative prices from other retailers or web-based suppliers. Do the retailer and consumer see this added useful complexity in the same way? Thinking back to the last chapters' discussions, what's the difference in view of the Ideal Result? Is there a way to make everyone happy? How could we add even more useful complexity to this service? Could we, after we have established the best price, go out to various credit card suppliers with my credit rating automatically supplied, to get the best payment terms? Lowest interest rate? Longest term "same as cash"? As long as the value of this added complexity pays for itself, it will be used and accepted. But what if the retailer did this all for you and demonstrated to you *in the store* that it had been done? Would you buy on the spot? If this were proven to you to be true after a few visits, what might happen to your loyalty? Would you leave your magic cell phone at home and not use it or the service anymore?

Now think about your home TV and entertainment systems. How many remotes do you have? Does the one that turns the TV on and off also do anything else on the systems connected to your TV? The complexity added to these systems may be useful in terms of the end results, but certainly not in the journey to get there. I suspect that many of you would pay significant money for a truly "universal" remote. We keep adding complexity to these devices, but it is questionable how useful these control devices are.

Our cars may be somewhere in the middle. In some high-priced cars, they have become moving entertainment systems—useful complexity. The unseen complexity involved in improving passenger and driver safety are also examples of added useful complexity. Some of this has been mandated, but it has reduced traffic fatalities and reduced insurance rates. The overall value of automobile travel is affected a great deal by fuel costs and the efficiency of the road system. Where we have discussed adding useful complexity, we've focused on what I'll call *local* alternatives (other office machines, additional oral care alternatives). When we think about cars and making them more useful and complex, we need to look at this as well. Is flying any quicker or less of a hassle? Are public transportation options available? What do they cost in terms of total cost and total time? What's the purpose of driving? To get somewhere? For what reason? To meet someone? Why not one of the virtual meeting services that is available over the Internet?

Do you remember when auto companies started incorporating a multitude of devices, monitors, and entertainment systems in their cars? We now have an issue of so many devices in the car that we worry about the driver becoming so distracted that actual driving has become a secondary concern, creating many safety concerns.

How can you decide whether adding useful complexity is the approach to take toward a more ideal product or service? Ask these questions:

1. Is what I am adding to the product or service going to replace something else? Does the new system cost less than the combined cost of the two products or systems? Have I managed to *trim* something, in total?
2. Is what I am adding to the product or system going to provide a more cost effective way to provide value to the customer without the customer paying?

Table 7.1 Analyzing added complexity

Function of your part or system	If we add this function:	This "other" product or system is eliminated

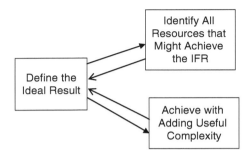

Fig. 7.2 The ideal result via resources and adding useful complexity

3. Is the added useful complexity going to enable the customer to do something they could not be done before?
4. Is the added useful complexity going to enable the customer to do something in multiple locations, under varying circumstances, that could not be done before?
5. Is the added useful complexity going to allow multiple users to do something though one system or device?
6. Has your customer ever asked you to "add" something to your product or service at no additional cost? What did you say? Why? Did you think long enough about why the question was being asked?

Table 7.1 can assist you.

Here's where we are in the process (Fig. 7.2).

As before, there is an interactive loop here, as when we discover how to add useful complexity, it may change our view of what the Ideal Result can or should be. It also might change our view of resource availability.

Chapter 8
Trimming: Another Approach to the Ideal Result

What is covered in this chapter is one of the most powerful of the individual TRIZ tools. It can be incorporated into *any* type of ideation or problem-solving process that doesn't necessarily involve TRIZ. It could be used as a part of any creativity session, independent of the structure or tools used. I am showing it where it is typically used in the process of using the entire TRIZ algorithm, but you could have a case or a situation, with limited time, and you just want to generate some quick breakthrough ideas and use it independently.

What comes to mind when you hear the word, "trimming"? Do you trim your hedges? Trim your payroll? Trim your budget? Have you ever "downsized" an organization? Been the victim of "downsizing"? Outsourced a function that traditionally been done in house? What are you trying to do? What was going on? When you trim your hedges, you are trying to improve the health of the shrubs, how your house looks, or possibly to remove dead limbs. When you trim your payroll, you are trying to save money, hopefully without harming your business. When you trim your budget, you are trying to save money, again without losing the value of what you aren't buying any more. Sometimes we know that's not going to be the case—we just want to save money and are willing to suffer the consequences.

In the TRIZ world, "trimming" is a formal tool where a disciplined analysis of a system or product components is made and then one of the parts of the system or process is deliberately (sometimes arbitrarily, sometimes focused on a high cost or high maintenance item) removed and the question asked, "How can the *function* of that now missing part, system, or process still be achieved with the remaining parts of the system or organization that are left?" This is very different than simply reducing manpower or supplies and telling everyone to figure out how to do more with less.

Now this may sound incredibly simple (and on the surface it is), but my experience is that the "trimming" concept is one of the most difficult breakthrough thinking concepts that exist. Our normal approach to analyzing a system is quite different. We look at a process or device with some kind of problem or a need to reduce operating cost. Engineers (especially!) look at this situation as an opportunity to design

J. Hipple, *The Ideal Result: What It Is and How to Achieve It*,
DOI 10.1007/978-1-4614-3707-9_8, © Springer Science+Business Media New York 2012

Table 8.1 Trimming

Part of a system to be eliminated	What does it do?	What else in the system could do this function?

something new to *add* to the system. While the problem may get solved, we have added complexity to the process or product. That addition, whatever it is, has to be monitored, controlled, maintained, depreciated, and eventually replaced. What if we approached this problem with a trimming mentality? We can get the same result, we save materials, maintenance, inventory, and cost.

Consider the recent announcements of the closing of some major newspapers and magazines, including the <u>Ann Arbor News</u> and <u>US News and World Report</u> as well as many investment letters. What is their function? To communicate information, including not only news but also coupons and shopping information. But we have a resource we didn't have decades ago—the Internet. People can get the news delivered at their computer screens and can print out what they want to spend reading in depth.

Several specialty computer manufacturers, including Light Blue Optics in England, are test marketing "optical" keyboards which project from the primary computer on to a flat surface, eliminating the need for traditional keyboard. When we "trim" the keyboard, we eliminate the use of plastics, cord and wire materials. Sound like the Pivoting RapidRoller™ we discussed earlier?

We can look at this from a number of perspectives. First, that of the computer system supplier. A new field (optical, electromagnetic) allows the elimination of the keyboard, cost of the plastics, shipping materials, and inventory space required. From the consumer's standpoint, less space and less complicated desk is needed. Now this invention is not without some issues—does it need a flat space vs. the ability to just put the keyboard on your lap? What about the loss of tactile feel of having hit a key properly? Is there a way to make sure that the desk surface is clear of paper and debris? Does dust interfere? But these are *secondary* problems. Remember what we said earlier? When a breakthrough idea occurs, we must isolate and list separately the secondary issues that we need to solve with other TRIZ tools and additional TRIZ analysis.

Now let's consider Apple's new tablet computer. No keyboard, no need for projection, no need for flat service, tactile feel is restored, etc. This device has trimmed so many external devices previously needed—a great example of trimming.

Here's a simple table to use in your "trimming" thinking (Table 8.1).

This last column in the table is sometimes, in some TRIZ literature, referred to as "feature transfer". We look for a way to use some other part of a product or system to accomplish the function that some other part of the system previously did and then eliminate that other system.

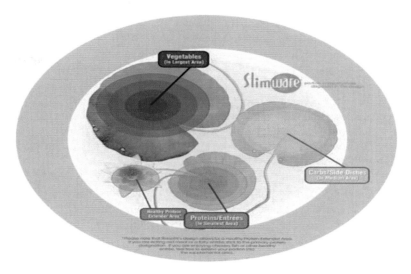

Fig. 8.1 The diet plate®

Let's take a look at some additional examples.

A new plate that has predesignated portion sizes already painted on the plate. A separate portion measurement is not required (Fig. 8.1).

Someone on a diet, or even more importantly, a diabetic patient, would find this "integrated plate" a very valuable product and several less dishes or containers to wash.

The Stanley Black & Decker Pivoting RapidRoller™ discussed earlier, "trims" the paint roller pan and the ladder. We'd like to eliminate the roller pan, but we still want its *function*. What resources do we have? If we place all three items on a table (the roller pan, the paint roller, and the amount of paint we need), and then asked that question, we might not come up with the answer that Stanley Black & Decker did. But what if we lay on the table *only* the paint roller and paint can and forced ourselves to get back the *function* of the roller pan that we once had? Would it be easier for you now to see that possibly there was no reason that the potential hollow in the stick handle was a resource you had not seen before? Trimming forces people to be inventive because we have taken away one or more resources they had. It's like doing product design on a desert island (the product development analogy to the Tom Hanks movie, "Castaway," where he made ingenious use of materials on a deserted island to survive for years before being rescued).The Michelin Tweel™, discussed previously, eliminated the normal tire and wheel cover in a wheel assembly.

Considering all the new tablet products, what has been trimmed out of the typical way we think of reading a book? What suppliers have been hurt? Benefitted?

If we held up a toothbrush and a tube of toothpaste and then *arbitrarily* took away the toothpaste tube and asked ourselves, "how can we get its function back with what's left?" What is the function that tube was providing? To hold and dispense toothpaste. But the toothpaste tube is now gone as illustrated in the freshngo™

toothbrush. Remember the Pivoting RapidRoller™? Why spend money on plastic to fill up the plastic handle when we could just as well fill it up with toothpaste? If you're the metal or plastic supplier to the toothpaste tube manufacturer, you've got a problem. Someone figured out how to get the *function* you provided with the resources they already had. Of course, they had to recognize the fact that an empty handle was a resource when previously it had always been filled. The *lack of space*, a void, is a resource! Another interesting thing to think about is how broad the technology search was by both Stanley Black & Decker and the developer of the toothpaste tube. Would either one look at the literature or patent literature of the other? That's what Altshuller did!

We have discussed medical and oral care several times already and let's revisit these two topics again in the context of "trimming." The "toothpaste in the handle" example just reviewed is a mechanical illustration of the trimming concept. Let's consider the service side of oral care. When was the last time your teeth were cleaned by the dentist (as opposed to a dental hygienist)? Hygienists only need to know a few of the skills that a dentist needs to know, they are paid less, and a dentist can spend time with more serious, higher income generating activities. This is an interesting example where the dentist has been "trimmed" from a task, but the office that the dentist owns has added useful complexity. Many medical doctors' offices have added "practitioners" of varying types, without full M.D. education, that can administer injections, do simple medical procedures, and provide "over the phone" assistance. Again, both trimming and adding useful complexity. In addition, the increasing role of government in paying for and controlling medical costs and procedures adds a political aspect to how the Ideal Result is defined. Certainly Republicans, Democrats, Libertarians, and Socialists do not define the Ideal Result in the same way.

The things that are "trimmed" that are acceptable will and can be greatly influenced by social, political, and other viewpoints. As with Ideal Result, resources, and adding useful complexity, these viewpoints need to be considered and analyzed. This is not a negative about applying TRIZ. It just makes it more challenging and interesting. Having group meetings where these differences are openly discussed and reviewed can provide an opportunity to define an Ideal Result that may be acceptable to all and means of using resources that is acceptable to all. Whether trimming or adding useful complexity is the correct path will *never* be clear cut. There will always be situations where adding useful complexity will be valuable to one person or group (meaning that they are willing to pay for it) and an unnecessary complication to another person or group. The pendulum of the market place will swing back and forth with time and there will always be some of both. From an innovation standpoint, it is important to *always* look at both sides of the curve and analyze the business and economic consequences and make an intelligent decision after having looked at both.

We have now added another box to our process (Fig. 8.2).

We've been through four steps in the process of getting to the Ideal Result. Any or all of them—envisioning the IFR, identification and use of resources, adding useful complexity, or trimming should have generated a number of ideas for your consideration both in problem solving and thinking about next generation products of

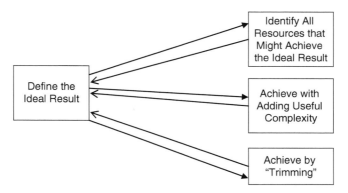

Fig. 8.2 Trimming added to the TRIZ diagram

services. Make a list of them and at the same time make notes about issues or problems you may see in implementation as well as information you may need to assess and further evaluate. DO NOT dismiss an idea at this point! Make a note about *why* an idea may not appear to be acceptable, *especially* if it appears to center around a contradiction.

It could be that one of the reasons you aren't excited about a particular idea is that you see a contradiction. Your thought process is probably framed by verbiage or thoughts such as "that's a good idea, but..." In many cases, this thought alone is enough to shut down further thought on an interesting new idea, especially if the comment is made by a senior manager in the room. In other cases, an idea is suggested by someone outside your technology area and you don't have enough information to assess, so you dismiss it. Don't do it! We'll discuss contradictions and their resolution next.

Exercises

1. Pick up the object closest to you right now. Where is it in the overall scheme of adding useful complexity to trimming? How did it get to the state it's in? What could be added to it that could be useful to you? Someone else? What could be "trimmed" from it that you could care less about? What is simplified in its manufacturing process? If a price reduction resulted, what would happen to product volume and shelf space? If we add some useful complexity, what higher end store would sell it? How much would the margin improve? What else could it be used for? In what new way could it be used?
2. What product or service do you provide to a customer that needs someone else's product or service to be useful to the ultimate customer? Are you tracking the patent and research activity of this "partner" to see what they are up to? Are they doing anything that might put you out of business?

3. Are you tracking the patent activity of your *customers* to see how they are think-ing about "trimming" you out of their product? What can you do to help them? Can you file a blocking patent that would cause a licensing discussion to occur? Could you jointly market the new concept?

4. Go back to Fig. 7.1 and plot every one of your products and/or services on the graph. Where are you? Why? If you think you're on the left side, what useful complexity can you add? If on the right side, what could you "trim" and simplify that would create a new product? If on the right side, how could you incorporate the "trimmed" functionality somewhere else in your product or service that would both save your money (we're not worrying about your supplier!) and delight your customer. We want both!

5. Your boss just came into your office and told you that 20 % of the staff in your organization was going to be eliminated and has asked for your help. How would you use TRIZ thinking to answer this question as opposed to seniority or perfor-mance ranking?

Chapter 9
Inventive Principles: What Do Millions of Patents Teach Us?

When someone says, "That's a good idea, but…" they are expressing a contradiction. Now if this is said by someone in authority at a meeting the idea may die just because of organizational barriers and TRIZ thinking can't solve that problem. But if it's an expression of frustration, we have a major opportunity. Remember the optimization graphs we looked at in Chap. 2? They are graphical representations of contradictions. If we try to go beyond a certain point in improving one parameter, the other parameter gets worse. We get so comfortable with this approach that we don't think about moving the curve (Fig. 9.1).

When Genrikh Altshuller, and those who followed in his steps, studied the breakthrough patents of the world (the top 5 % or so) they found one fundamental characteristic of virtually all of them: *They resolved a difficult contradiction. Difficult* contradictions! (I use the term "Excedrin® headache" contradiction in my work with clients to distinguish them from normal engineering design contradictions that may be relatively straightforward to resolve). They provide a mechanism or process allowing the resolution of contradictions. In the study of these breakthrough patents, an amazing fact was discovered which still holds true today, and that is that there are basically 40 high-level inventive principles that were used in any of these patents. As study of the patent literature continued, updating changes in businesses and technologies, we still find that a huge percentage of inventions and patents use these same 40 inventive principles. I am only using the word "huge" because I haven't read all the patents in the world, but I can say that I personally have not seen one that cannot be mapped against one of these principles with the exception of some chemical and biological mechanisms. Before we look at the contradiction table itself, let's review these 40 principles and see examples of their use in inventions that you will recognize. This is the opposite of how these principles are reviewed in other TRIZ publications. Most other TRIZ experts start with a review of the TRIZ contradiction table and then cover the 40 principles. I have chosen to reverse this introduction since the 40 principles can be used independently of the table. Simply throwing them out randomly in a brainstorming session can improve the productivity of such a session since the stimulation is coming from a demonstrated list of inventive principles based on the world's

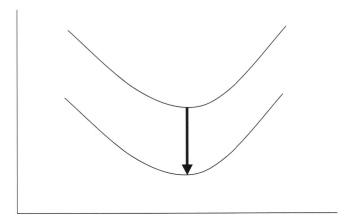

Fig. 9.1 Moving the optimization curve

most significant inventions. This is superior to the use of random words, pictures, objects, or music. It is common to hear creativity leaders and facilitators whose primary focus is psychology tell their groups that their success is measured and judged by the quantity of ideas generated. The primary explanation given is that, since so few ideas turn out to be worthwhile, the goal should be to generate as many ideas as possible since so few will actually be useful. TRIZ takes the opposite approach. It says basically, since there are only 40 principles that have solved any serious problem or challenge over a 60+ year period, there is little point in generating additional ideas that have shown little or no value. The key is to adequately define the problem so that we can match the problem with a known solution. This is not the most productive use of these principles, but it is a low level use and can assist in the integration of TRIZ with other creativity and problem-solving methods, if desired, or introduce TRIZ in a nonthreatening fashion which then might produce additional curiosity.

This list of principles, regardless of how they are published and summarized, has maintained its initial numbering, but suffered to some extent from minor language differences in descriptions of the same principle. For example, one table might say "nesting," while another might say "Matroiska." If you are of eastern European ancestry, you know what this is because you have seen Matroiska dolls that nest within each other. We will try to summarize some of these differences, but it is possible that you may see a particular inventive principle described with different language, but if it labeled inventive principle #17, it is referring to the same principle described in this chapter or that in other TRIZ publications. Some TRIZ software vendors and researchers have subdivided these principles and will claim that there are 70, 160, or as many as 400 principle, so make sure that when discussing TRIZ inventive principles with someone that you are having an "apples and apples" discussion. For the most part, these additional principles are subdivisions of the higher level 40 principles. They may be of some assistance in stimulating ideas but are not necessary in most cases.

After we review the principles, we will review the TRIZ contradiction table and how the inventive principles relate to it. Later on, we will discuss the priority of use of these principles, again as derived from the study of the global patent literature. This information can also be used to increase the productivity of the use of the 40 principles as part of an alternative process. Again, this is not the recommended use of these principles, but my experience has taught me that many people will make use of only selective portions of the TRIZ algorithm and tool kit and I want to make it as easy as possible for the water to be sampled in the hope that this will increase the thirst for the rest of what is in the reservoir!

The TRIZ 40 inventive principles are listed in Table 9.1.

The numbering of the principles is uniform in any TRIZ publication.

Let's go through the 40 principles one by one, and with each, provide ten examples to help illustrate the use of the principle in many different areas.

These lists are a composite of lists of my own [1, 2], those of Ellen Domb [3], and those of Darrell Mann [4].

Table 9.1 The TRIZ 40 inventive principles

1. Segmentation	15. Dynamics	29. Pneumatics and hydraulics
2. Taking out	16. Partial or excessive action	30. Flexible shells and thin films
3. Local quality	17. Another dimension	31. Porous materials
4. Asymmetry	18. Mechanical vibration	32. Color changes
5. Merging	19. Periodic action	33. Homogeneity
6. Universality	20. Continuity of useful action	34. Discarding and recovering
7. "Nested Doll"	21. Skipping (through)	35. Parameter change
8. Anti-weight	22. "Blessing in Disguise"	36. Phase transitions
9. Preliminary anti-action	23. Feedback	37. Thermal expansion
10. Preliminary action	24. "Intermediary"	38. Strong oxidants
11. Beforehand cushioning	25. Self-service	39. Inert atmosphere
12. Equipotentiality	26. Copying	40. Composite materials
13. "The Other Way Around" ("Do It In Reverse")	27. Cheap, short living objects	
14. Spheroidality/curvature	28. Mechanics substitution	

Inventive Principle #1: Segmentation

Segmentation means dividing a system or object into independent parts or increasing the degree of segmentation of an object or system.
Examples of the Use of This Principle:

1. Dividing a complex project or process into parts that are dealt with, analyzed, or planned by special resources or focus
2. Modularizing a product into parts and segments that make it easy to take apart or assemble
3. Divide marketing efforts according to market segments, sales force skills, and seasonality
4. Mass customization in product assembly (original Dell Computer business model)
5. Coupons printed on the back of sales receipts based on products purchased
6. Separation of food products within a larger package (Kraft Lunchables™)
7. Separate food packages such as cheese slices, and cereal packs
8. Closet separators
9. Segmented window covering such as Venetian blinds
10. Quick disconnects in mechanical systems

Where have you used/seen used this principle?

_____

_____

_____

Inventive Principle #2: "Taking Out"/Trimming/Physical Separation

"Taking out" refers to the removal of a part of a system, product, or process which may be unnecessary, may provide some advantage, or minimize a disadvantage.
Examples of the Use of This Principle:

1. Isolating a noise producing part of a mechanical system to minimize the need for noise protection
2. Outsourcing a "non-core" competence to a contractor
3. Providing a separate shopping location for a particular market segment (kids vs. adults)
4. Separating and isolating the storage of certain chemicals with certain safety or physical properties
5. Selling base products with few complicated features which can be added on at a later date at added cost
6. Removing an ingredient in a food product that has no verifiable benefit but has

been incorporated out of tradition

7. Use of independent interviewers and surveys to isolate the information seeker from the answers to insure independence
8. Allowing use of air conditioning cut off due to high demand in exchange for lower rates
9. Use of artificial landscaping products where live plants and grass use high levels of water and fertilizer
10. Closing off vents in a house in rooms not in use

Where have you used/seen used this principle?

_____
_____
_____

Inventive Principle #3: Local Quality

Changing the structure of a product, system, or process from uniform to nonuniform. Make a product, system, or process responsive to a particular "local" situation or condition. Make different parts of a product, process, or service perform different functions.

Examples of the Use of This Principle:

1. Varying volatility and octane specifications of gasoline as a function of seasonality, local conditions, or usage condition
2. Change pricing as a function of local conditions, situations, or weather
3. Customized packaging and marketing as a function of geography or retail conditions
4. Varying the properties of a tape or film on its surface versus its bulk properties
5. Targeted radio pharmaceuticals that go to a particular organ in the body
6. Use of focus groups in marketing research that gather data from particular groups as a function of age, ethnicity, sex, or experience
7. Multiple function tools
8. Multiple product functionality such as different SPF designations in sunscreens
9. Coatings on food products or coatings for preservation
10. Different packaging shapes and sizes based on consumer need (travel size products)

Where have you used/seen used this principle?

_____
_____
_____

Inventive Principle #4: Asymmetry

This principle refers to the use of nonuniformity, primarily in a geometrical sense, but could also mean in a time or sequence sense.
Examples of the Use of This Principle:

1. Asymmetric design of a the Boeing 737 jet engine to allow necessary ground clearance as the engine size is increased
2. Use of asymmetric tank mixing systems to increase mixing uniformity
3. Left and right hand biased physical devices
4. Use of dissonance in music composition to increase intensity
5. Nonuniform organizational structure that relates to customer rather than supplier human resources department ideas
6. Architectural design to catch the eye of an observer
7. Irregular shaped signs to catch the attention of an observer
8. Mattresses with different firmness as a function of location
9. Use of wave or noise cancellation technology
10. Use of multilayer products that provide different functionality at the surface versus the bulk of the product

Where have you used/seen used this principle?

_____

_____

_____

Inventive Principle #5: Merging/Combining

This principle refers to the combining and/or merging of two or more functions, typically done independently, into one system, process, or product.
Examples of the Use of This Principle:

1. Merging of office equipment into one device (copier/fax/scanner, etc.)
2. Networked versions of computers and other devices
3. Use of networks of self-employed individuals versus full time staff
4. Integration of summer and winter consumer services such as lawn care and snow removal; mulching lawnmower
5. Swiss Army knives
6. Corporate mergers to broaden capabilities; "one stop" shopping
7. Multifunctional tools (can/bottle opener)
8. Dual paint rollers to apply two separate colors simultaneously
9. Parallel computer processing; parallel processing in new product development
10. Debt consolidation

Where have you used/seen used this principle?

_____

_____

_____

Inventive Principle #6: Universality

This principle refers to making a product, system, or process useful in multiple applications,... environments or under varying conditions. Use of this principle frequently eliminates another component or separate system/product.
Examples of the Use of This Principle:

1. An adjustable wrench (also an example of principle #4, dynamism)
2. "Matrix" management structures when managers and employees have cross functional organizational structures with "dotted lines"
3. Consultancy across different areas and different business cards to represent each
4. Reconfiguration of airplane seating arrangements to reflect airfare incentives, type of airplane, or other commercial factors
5. Management structure that varies with the nature of the external customer
6. Ability of a computer to recognize and adapt to the type of document being processed
7. Design of a PC to use multiple input electrical voltages
8. Airport jetway design to accommodate multiple types of airplanes
9. Copiers that automatically adjust settings based on the characteristics of the original document
10. Combination normal and Phillips screwdriver heads

Where have you used/seen used this principle?

_____
_____
_____

Inventive Principle #7: "Nested Doll"

This principle refers to the concept of components or parts of a system being "nested" within a supersystem to conserve space, improve the use of unused volume, or containing one thing inside another in a controlled way. As mentioned previously, an interesting example Eastern European or Russian example of this principle is a Matroiska doll. This can also refer to parts of a system passing through cavities within the system.
Examples of the Use of This Principle:

1. Design of shipping containers to contain smaller units inside of larger ones
2. A business structure where small companies are structured under a larger corporate umbrella
3. Nesting within graphic displays
4. Nesting within software architecture
5. Nesting within process control algorithms

  6. Flexible pointers
  7. Zoom lenses
  8. Seat belt retraction mechanism designs
  9. Retractable wheels in airplanes
10. Measuring cups/spoons of varying sizes that nest within each other

Where have you used/seen used this principle?

_____
_____
_____

Inventive Principle #8: Anti-weight

This principle refers to the use of weight or other property compensation by internal system design or interaction with the system around the product or system.
Examples of the Use of This Principle:

  1. Use of center of gravity close to the sagittal plane of a body to make lifting and carrying easier
  2. Use of foams or lightweight gases (helium) to provide buoyancy
  3. Hydrofoils
  4. Use of short sales to counter stock market euphoria; hedging
  5. Use of countering viewpoints and personalities in assembling teams and advisors
  6. Use of teachers' aides or paralegals to compensate for lack of time
  7. Use of pressure or vacuum to drive unequal chemical stoichiometric gas reactions in a favorable way
  8. Eductors and vacuum jets in fluid handling systems
  9. Aircraft wing design to provide lift
10. Assembly line counterweights

Where have you used/seen used this principle?

_____
_____
_____

Inventive Principle #9: Preliminary Anti-action

This principle refers to the use of a process or system design to prevent a possible future negative event, result, or harmful effect.
Examples of the Use of This Principle:

1. Chemical solution buffering
2. Prestressed concrete
3. Sunscreen lotions
4. Tape products to prevent paint leaking on to the wrong surface
5. "Poison pill" to prevent unwanted corporate takeover
6. Polymerization inhibitors
7. Dead man switches; safety interlocks
8. Pretension to avoid kickbacks in machinery
9. "Straddle" stock purchases
10. Boot camps, zero gravity exposures prior to real experience

Where have you used/seen used this principle?

_____

_____

_____

Inventive Principle #10: Preliminary Action ("Do It in Advance")

This principle refers to performing a function before it is needed, prearranging processes or products or preparing a system for change
Examples of the Use of This Principle:

1. Job planning to avoid accidents and potential injuries
2. Employee training; emergency drills
3. Preformatting documents
4. "Stop loss" stock market orders
5. A legal will
6. Establish a line of credit before a major purchase or acquisition
7. Prepasted wall paper
8. Evacuation of major metropolitan areas prior to natural disaster such as a hurricane
9. Preheating and precooling
10. Storage for buffering of production rates with intermediate tankage and storage

Where have you used/seen used this principle?

_____

_____

_____

Inventive Principle #11: Beforehand Cushioning

This principle refers to the incorporation of features that cushion or mediate a surge or sudden change, especially one that is considered to be likely
Examples of the Use of This Principle:

1. pH buffers in solution
2. Air bags in cars
3. Emergency backup systems
4. Security for a loan
5. Call or put option
6. Insurance policies
7. Severance contract
8. Lockout systems
9. Confirmation dialogue box in software or programs
10. Prenuptial marriage agreement

Where have you used/seen used this principle?

_____

_____

_____

Inventive Principle #12: Equipotentiality

This principles refers to equalizing fields and energy when needed, primarily by advance system or process design.
Examples of the Use of This Principle:

1. Canal locks
2. Limiting possible position changes in a physical system
3. Spring loaded parts
4. Gravity fed soda dispenser
5. Self-leveling devices
6. Organizing educational classes by pretests, etc.
7. Teams deciding how to distribute special awards for achievement vs. senior management

8. Team assignments with differing points of view to insure all sides of an issue are given equal time
9. Cable car weight balancing
10. Mechanic's pit instead of raising a car for maintenance

Where have you used/seen used this principle?

_____

_____

_____

Inventive Principle #13: "Other Way Around"/Do It in Reverse

This principle refers to reversing the normal procedure or process.
Examples of the Use of This Principle:

1. Moving sidewalk
2. To loosen a part, cool the inside versus heating the outside
3. Treadmill (you don't move linearly but the treadmill does)
4. Reverse order of chemical mixing to minimize hazards of heats of solution (sulfuric acid and water)
5. Rotate part instead of the tool
6. Home visits by medical personnel; delivery of "meals on wheels"
7. Customers find suppliers via the Internet versus direct sales calls
8. Evaluation of managers by subordinates
9. Wind tunnels (vehicle stands still and the wind moves)
10. Home shopping

Where have you used/seen used this principle?

_____

_____

_____

Inventive Principle #14: Curvature/Spheroidality

This principle refers to the use of curvature and spherical geometry (and thinking!) vs. straight line and linear. Move from linear to rotary motion. Use of centrifugal force. Cyclic planning and actions.
Examples of the Use of This Principle:

1. Ergonomically shaped keyboards
2. Roller balls, ball point pens, and spherical castors
3. Screws versus nails

4. Centrifugal separators (Dyson® vacuum cleaners), spin cycles in washing machines
5. Cyclone separators
6. Spherical versus linear exercise machines
7. Lens curvature to improve visibility
8. Arches and domes for improved structural strength
9. Rotary actuators
10. Revolving credit agreements

Where have you used/seen used this principle?

_____
_____
_____

Inventive Principle #15: Dynamics/Dynamism

This principles refers to the introduction of dynamism and responsiveness into a product, system, or process.
Examples of the Use of This Principle:

1. Price versus quality, volume, delivery speed, etc.
2. Variable speed transmissions
3. Variable pitch screw conveyors
4. Flexible sigmoid scope
5. Variable speed pumps
6. Changing organizational structure based on business conditions or competitive situations
7. Cafeteria benefit plans based on needs, age, family size, and cost
8. Shape memory materials
9. Flexible joints, straws, and piping
10. Process control settings that vary with product desired

Where have you used/seen used this principle?

_____
_____
_____

Inventive Principle #16: Partial or Excessive Action

Use of less or more of the originally desired action, effort, or field when exactly the right amount is hard to achieve.
Examples of the Use of This Principle:

1. Overfilling of holes and then sanding or finishing down
2. Use of the 80/20 rule to focus on the most important actions necessary
3. Communicate more often than necessary to insure message is heard
4. "Over spraying" to make sure coverage is complete
5. Over design in engineering fields to insure system does not fail, overpaying on mortgage to build up equity faster
6. Rent rather than buy
7. Excess reactants in a chemical reaction to increase conversion
8. Flooding a process with water to kill a hazardous reaction
9. Stretch goals for employees and bonus decision
10. Purchasing the minimum amount of home insurance to provide coverage

Where have you used/seen used this principle?

_____

_____

_____

_____

Inventive Principle #17: Another Dimension

Move to two or three dimensional space as opposed to a one or linear dimension; reorientation of an object.
Examples of the Use of This Principle:

1. Holograms
2. Multi-story versus single story housing, parking garages, shelves, and structures
3. Multidimensional computer chip architecture
4. Multiaxis cutting tools
5. Dump trucks
6. Multilevel CD carriages
7. Braille writing for the blind
8. 3-D imaging in the medical and security areas

9. Traditional hierarchical management structure
10. Use of internal pores of a solid for reaction surfaces and sites

Where have you used/seen used this principle?

_____

_____

_____

Inventive Principle #18: Mechanical Vibration

Cause an object to oscillate or vibrate. Increase the frequency of oscillation or vibration. Use an object's resonance frequency.
Examples of the Use of This Principle:

1. Ultrasonic flow measurement, cleaning, and drying systems
2. Use of vibration to dislodge frozen or adhered solids from storage systems
3. Tactile feedback
4. Helmholtz and piezoelectric resonators
5. Rough road "shoulder" to alert dozing driver; rumble strips
6. Gall and kidney stone destruction
7. Quartz crystals for high accuracy clocks
8. High frequency dog whistles
9. Vibrating carving knives
10. Vibrating cell phones versus noisy signal

Where have you used/seen used this principle?

_____

_____

_____

Inventive Principle #19: Periodic Action

Replace continuous action with pulsed or periodic action. If action is already periodic, change the frequency or magnitude to improve functionality. Use gaps in action to perform useful actions or functions.
Examples of the Use of This Principle:

1. Batch manufacturing to improve quality and isolation
2. Emergency training and drills during down times
3. Rotation assignments
4. Printing during carriage returns
5. Multiskill and cross-training to make use of all human resources at all times
6. Random checking for drugs, safety, and hazardous materials

7. Pulsed versus continuous sounds for attention; pulsed showers
8. Variable speed pumps versus continuous pumps and throttle valves
9. Pulsation to produce fine particles
10. ABS brakes

Where have you used/seen used this principle?

_____

_____

_____

Inventive Principle #20: Continuity of Useful Action

Carry on work or process continuously, eliminate idle time, and work at full load and optimum efficiency at all times.
Examples of the Use of This Principle:

1. Flywheels and capacitors to store energy
2. Constant pitch/variable speed propeller
3. 24 h store operations
4. Practice while not competing
5. Cross-training during idle times
6. Self-cleaning systems
7. Auto-defrost
8. Heart pacemaker
9. Continuous computer back up
10. Double ended canoe or kayak paddles

Where have you used/seen used this principle?

_____

_____

_____

Inventive Principle #21: Skipping/Rushing Through

Conduct a hazardous or costly operation at high speed to minimize harmful effects.
Examples of the Use of This Principle:

1. High speed dentists' drilling
2. Flash photography
3. Cutting materials faster than heat can propagate
4. Pay off loan early

5. Accelerated depreciation
6. Minimize reaction times under known hazardous conditions
7. Bypass known flammable regions for chemical handling
8. "Fail fast" in the prototyping stage of a new product
9. Minimize intermediate storage of hazardous chemical reactants
10. Rapid firing of employees to minimize useless discussions

Where have you used/seen used this principle?

_____
_____
_____

Inventive Principle #22: Blessing in Disguise/"Lemons into Lemonade"

Use harmful factors, conditions, or situations in a positive way. Add harmful factors together to produce a positive result. Amplify a harmful factor to such a degree that it is no longer harmful. Use negative as a positive
Examples of the Use of This Principle:

1. Using byproduct waste energy to preheat feedstocks, generate electricity (cogeneration), recharge batteries in cars (Toyota Prius™)
2. Collection of negative customer feedback in real time to allow immediate correction (Internet feedback)
3. Use of a backfire to eliminate fuel from a forest fire use of explosives to put out oil well fire cause corrosion chemistry to create a protective layer of corrosion products
4. Short selling in the stock market
5. Ask "complainers" within a team to produce an action plan to prevent negative concerns put managers who complain about performance of another department in charge of it
6. Couple two totally opposite personalities on a project
7. Vaccinations
8. Prewashed denim jeans look worn, but sell for more money
9. "Loss leader" to bring people into a store to buy other things
10. Reduce resources to force people to find new ways to accomplish goals false scarcity to create novel solutions

Where have you seen this principle used?

_____
_____
_____

Inventive Principle #23: Use of Feedback

Introduce feedback. If feedback is already used, change its magnitude or influence.
Examples of the Use of This Principle:

1. Closed loop process control with varying algorithms; automatically changing as a function of process demands
2. Store loyalty cards and rewards
3. Cause interaction between designers and customers
4. Employee feedback to senior manager
5. Car radio volume changes automatically as car speed changes
6. Statistical process control
7. Automobile fuel combustion control via end of pipe measurements
8. Gyrocompass controls autopilot
9. Changing color of signal to signify seriousness
10. Automatic ordering from suppliers based on cash register inputs

Where have you seen this principle used?

_____

_____

_____

Inventive Principle #24: "Intermediary"

Use an intermediate between two objects, systems, or people. Use an intermediate which is easily removed after needed function is completed.
Examples of the Use of This Principle:

1. Drink coasters
2. Insulating gloves from heat or cold
3. Use of catalysts in chemical synthesis that are later separated
4. Carpenter's nail set
5. Hub city for airlines
6. Use of mediation board in labor relations, ad hoc committees
7. Pop-up windows in software programs under certain conditions
8. Paper clip
9. Escrow agents, leasing agents, and brokers
10. Bridge loan

Where have you seen this principle used?

_____

_____

_____

Inventive Principle #25: Self-service

Make an event, product, or service perform additional auxiliary useful functions.
Use waste resources, energy, and materials.
Examples of the Use of This Principle:

1. Self-serve gasoline stations, fast food restaurants, and ATM's
2. Faucet water purification
3. Waste heat recovery, and geothermal energy use
4. Prius™ waste brake heat to recharge battery
5. Autocatalytic chemical reactions, and self-cleaning oven
6. Cogeneration of power and steam in the chemical industry
7. Cell phones for airline check-in and bank deposits
8. Self-help groups
9. Building materials that scrub pollution gases, lawn waste as fertilizer; composting
10. Optical illusions used deliberately

Where have you seen this principle used?

_____

_____

_____

Inventive Principle #26: Copying

Use cheap, disposable substitute for expensive, fragile, or difficult to replace object.
Replace object with optical copy. Use infrared or UV copies versus optical copies.
Examples of the Use of This Principle:

1. Virtual reality, crash test dummy
2. Listen to recordings (or attend via Internet) versus attending live performance; prerecorded announcements
3. Infrared imaging to detect hot spots, diseases, intruders
4. Survey from space and satellites measurements via photos
5. Artificial grass material, imitation jewelry, color prints of original paintings, store brands versus name brands
6. "Fake" mold containing lunch bags (http://www.thinkofthe.com/)
7. Sonograms to evaluate health of fetus versus direct measurement
8. Video conferencing versus travel
9. Cloud computing and virtual software programs
10. Use of benchmarking findings to improve strategy

Fig. 9.2 The Moldy sandwich bag (http://www.thinkofthe.com)

Here is a picture of example #6. A company has created a sandwich bag with a preprinted picture of mold on it so that fellow employees are less likely to steal others' food from common refrigerators (Fig. 9.2).
Where have you seen this principle used?

_____

_____

_____

Inventive Principle #27: Cheap Short-Living Object

Replacement of an expensive object, system, or process with a multitude of inexpensive, short life objects, systems, or processes.
Examples of the Use of This Principle:

1. Disposable dishes, glasses, silverware
2. Disposable lighters and diapers
3. Temporary traffic rerouting barriers
4. Temporary employees, rehiring of retired people on a short-term basis
5. Inexpensive chemical catalysts that stay with products
6. Constant removal and replacement of filter aids in centrifugal filters

7. Use models for operator and maintenance training
8. Virtual war games, simulators
9. Bridge loans and options trading
10. Sacrificial coatings

Where have you seen this principle used?

_____
_____
_____

Inventive Principle #28: Mechanics Substitution

Replace a mechanical system, force, or system with one based on optical, acoustic, electronic, or chemical field. Use of electric or magnetic fields to enhance a mechanical field. Change from static to movable and structured fields. Use substances that enhance the use of nonmechanical fields.
Examples of the Use of This Principle:

1. Acoustical field to replace physical fences for pets
2. Electro-rheological fluids magnetic bearings
3. Photochromic glasses
4. Adding odorant to detect presence of a material; natural gas odorizing with mercaptans
5. Opposite charge materials to induce mixing
6. Ferromagnetic materials
7. Touch screen entry versus hand completed forms
8. GPS locators
9. TV and audio remote controls
10. Voice-activated commands

Where have you seen this principle used?

_____
_____
_____

Inventive Principle #29: Pneumatics and Hydraulics

Use gas and liquid parts instead of solid objects and systems
Examples of the Use of This Principle:

1. Inflatable mattresses and furniture
2. Gas bearings
3. Gel inserts for shoes
4. Use of hydraulic systems to store energy
5. Pneumatics and hydraulics to reduce necessary human force requirements
6. Flexible management structure
7. Converting from illiquid to liquid assets
8. Pneumatic conveying
9. Pneumatic control systems
10. Load cells

Where have you seen this principle used?

_____
_____
_____

Inventive Principle #30: Flexible Shells and Thin Films

Flexible films and structures versus rigid 3-D structures. Isolation of systems and products via thin films or shells. Flexible systems.
Examples of the Use of This Principle:

1. Plate and frame heat exchangers
2. Gas separation and water desalination membranes
3. Tea bags
4. Bubble wrap
5. Inflatable fake passengers for toll-free lanes
6. Spam filters
7. Prepackaged materials
8. Protective covers and films
9. Revolving lines of credit
10. Honeycomb or "buckyball" chemistry structures

Where have you seen this principle used?

_____
_____
_____

Inventive Principle #31: Porous Materials

Make an object or system porous. Increase porosity to introduce a positive or useful substance or function.
Examples of the Use of This Principle:

1. Holes to reduce weight
2. Foamed materials, sponges
3. Medicated swabs and dressings
4. Embedded sensors into cavities
5. Encourage open mindedness and receptivity to new ideas
6. Storage of hydrogen in palladium sponge material
7. Porosity design of catalysts to control selectivity based on molecule size, automobile catalytic converters
8. Deliberate "leaks" to test public or enemy reaction
9. Wicking in clothing design to provide cooling
10. Controlled information flow

Where have you seen this principle used?

_____

_____

_____

Inventive Principle #32: Color Changes

Change the color of a product, object, or surroundings. Change the transparency of an object or system. Change emissive properties subject to radiation.
Examples of the Use of This Principle:

1. Use color to indicate degree of hazard in traffic, chemical systems, pH paper
2. Color sensitive labeling (e.g., food spoilage temperature)
3. "Six Thinking Hats®" process to control structure of idea generation
4. Highlighters
5. Electrochromic glass
6. Fluorescent additives
7. Parabolic receptors
8. Photolithography
9. Camouflage
10. Safe lights in photographic darkrooms

Where have you seen this principle used?

_____

_____

_____

Inventive Principle #33: Homogeneity

Make objects interact with a given object of the same material or material with similar properties.
Examples of the Use of This Principle:

1. Make ice cubes out of the drink into which they are going to be used
2. Compostable plant pots
3. Use materials for structures that have similar coefficients of expansion and conductivity to minimize structural shape changes
4. Minimize differences in electrochemical potential to minimize corrosion
5. Provide uniform customer experiences independent of location
6. Use local currency
7. Common goals that support corporate strategy
8. Use identical blood types for transfusions
9. Wooden dowel joints
10. Affinity mapping

Where have you seen this principle used?

_____
_____
_____

Inventive Principle #34: Discarding and Recovering

Make portions of an object or system go away after they have performed their useful function. Restoration of functionality during normal operation.
Examples of the Use of This Principle:

1. Dissolving capsules for medicine
2. Use of dry ice as a sand blasting agent
3. Temporary or seasonal employees
4. Self-sharpening blades
5. Self-tuning engines
6. Packaging of needed use sizes in disposable packaging
7. Autocatalytic chemical reactions
8. Sub-contracting sudden work surges
9. Warranty guarantees
10. Filters which are cleaned and reused

Where have you used/seen this principle used?

_____
_____
_____

Inventive Principle #35: Parameter Change

Change an object's physical state (solid/liquid/gas). Change concentration or consistency. Change of degree of flexibility. Change the temperature. Change pressure or other physical parameters.
Examples of the Use of This Principle:

1. Controlled explosion in automobile air bags
2. Font, case, and italics use
3. Reaction time based on how information is presented
4. Use of Curie point
5. Heat or cool for rigidity or flexibility
6. Liquids versus powder detergents
7. Dry ice
8. Gelled materials, jellies
9. Pressure cookers, raising pressure or reducing pressure to change boiling points
10. Thixotropic fluids

Where have you seen this principle used?

_____
_____
_____

Inventive Principle #36: Phase Transitions

Proactive use of phase transitions (volume or other physical property phenomena).
Examples of the Use of This Principle:

1. Use of latent heats of freezing and boiling
2. Superconductivity at very low temperatures
3. Water expansion when freezing (different than most other liquids)
4. Digital versus analog
5. Film to DVD conversion
6. Changes in business conditions
7. Use of heat pumps
8. Use of expansion and contraction during freezing and melting
9. Buy/sell receivables to obtain cash
10. Storage of energy in phase change material

Where have you seen this principle used?

_____
_____
_____

Inventive Principle #37: Thermal Expansion

Use thermal expansion or contraction in a positive way. Use variation in thermal properties to achieve useful effects.
Examples of the Use of This Principle:

1. Shrink wrapping
2. Expansion joints
3. Bimetallic strips for thermometers, bi-metallic hinges for self-opening blinds
4. Shape memory alloys
5. Self-regulating organization, differing stimulus for individuals and teams
6. Arbitrage
7. Insurance premium changes with perceived dangers
8. Leaf spring thermostats
9. Marketing efforts respond to popularity of product
10. Thermal switch cutouts

Where have you seen this principle used?

_____
_____
_____

Inventive Principle #38: Strong Oxidants

Use of oxygen-enriched air, pure oxygen, or ozone. Use of ionizing radiation or ionized oxygen. Use of focused energy (of a technical or human nature) of any sort.
Examples of the Use of This Principle:

1. Hospital oxygen tents, localized oxygen breathing equipment
2. Use of nitrous oxide
3. Irradiation for food preservation
4. Ozone to destroy harmful organisms
5. Focused audits of organizations
6. Oxidizing cleaners
7. High risk, high return; high risk to stimulate effort
8. Actual case studies in training and education
9. Keynote speakers, inject "new blood" into teams, devil's advocates
10. Accelerated oxidation reactions

Where have you seen this principle used?

_____
_____
_____

Inventive Principle #39: Inert Atmosphere

Replace normal atmosphere with an inert one. Add neutral or inert elements.
Examples of the Use of This Principle:

1. Use of argon in welding
2. CO_2 fire extinguisher, foam extinguishers
3. Vacuum packaging
4. Use of dampers
5. Sound absorbing panels
6. Helium filled high insulation glass windows
7. Aggregates in concrete
8. Add "neutral" individuals to teams, discussions, arbitration panels
9. Inert ingredients to bulk up food products, use of inert pipeline fillers to separate fluid segments
10. Meetings in neutral, retreat locations; rest breaks

Where have you seen this principle used?

_____

_____

_____

Inventive Principle #40: Composite Materials

Change from uniform to composite materials.
Examples of the Use of This Principle:

1. Concrete aggregate
2. Fiber reinforced plastics and ceramics
3. Composite golf clubs, aircraft materials
4. Nontraditional work structures and teams
5. Proactively using different personality styles on teams
6. Multilayer films and membranes for fluid and gas separation
7. Loan with a balloon payment
8. Training that uses many different styles of delivery
9. Heterogeneous workforce based on training and discipline
10. Carpet structures

Where have you seen this principle used?

_____

_____

_____

Come up with at least one illustration of each principle to ensure you understand its application and can explain it to someone else. In addition, consider how you might use that principle in your own product, business, or organization. You should have at least two entries under each inventive principle for a total of 80 ideas.

Frequency of Principle Use

In the study of the patent literature in the context of the TRIZ 40 principles (www.triz-journal.com/archives/2004/04/01.pdf), it has also been possible to "force rank" how often a particular inventive principle is used in the patent literature. It is also not surprising that this frequency has changed as business and technology has changed significantly over the 70 year period after the original TRIZ development to today's complex set of tools including problem modeling software and the linkage to the current patent data base. The original study of the patent literature showed the frequency of use of the 40 principles (Numbers are the frequency of use not the principle numbers) (Table 9.2).

In an updated study conducted by Mann (www.triz-journal.com/archives/2004/04/01.pdf), use frequency has changed. The degree to which their frequency/less frequency of use if also highlighted in terms of how far they have advanced up the list (Again, remember that these numbers refer to frequency of use, *not* the numbers of the principles). Reasons for these changes will be discussed next (Table 9.3).

Before we look at the principles that have increased or decreased in the frequency of use, look at which parameter is still #1, after 60 years! *Parameter change*—why? It is because the change in a system's basic parameters usually causes fundamental change in properties and/or a large energy change, both of which are potential source of significant change to a system or product.

Let's consider some possible reasons for the *increased* use of certain principles.

Merging—what has happened to all the separate office machines you used to have in your office? How many devices are now incorporated into your TV, car, and alarm clock or radio? Advances in electronics have allowed us to combine many functionalities into a single device. In your processes or products, how could you combine functions into fewer pieces and components?

Equipotentiality—In virtually all mechanical assembly lines we see the use of springs, leveling and positioning devices to bring needed parts to a position where the operator or technician does not have to stretch or reach, minimizing ergonomic injuries.

"Nested doll"—Space is precious and any system which allows the use of previously unused space is in favor. Extended pointers, zoom lenses, and radio antennas are all examples of the use of this principle. The nesting of small businesses under a larger corporate umbrella is another. Many companies, when they acquire another company, keep the original company's trade names, logos, etc. to preserve the

Table 9.2 Original frequency of use of TRIZ inventive principles

1. Parameter change	15. Discarding and recovering	29. Before-hand cushioning
2. Preliminary action	16. Partial or excessive action	30. Porous materials
3. Segmentation	17. Composite materials	31. Strong oxidants
4. Mechanics substitution	18. "Intermediary"	32. Anti-weight
5. Taking out/separation	19. Another dimension	33. Merging
6. Dynamization	20. Universality	34. "Nested Doll"
7. Periodic action	21. Curvature/spheroidality	35. Skipping
8. Mechanical vibration	22. "Blessing in Disguise"	36. Feedback
9. Color change	23. Inert atmosphere	37. Equipotentiality
10. "The other Way Around"	24. Asymmetry	38. Homogeneity
11. Copying	25. Flexible shells/thin films	39. Preliminary anti-action
12. Local quality	26. Thermal expansion	40. Continuity of useful action
13. Cheap, short-living objects	27. Phase transitions	
14. Pneumatics and hydraulics	28. Self-service	

Table 9.3 Change in frequency of use of the 40 principles

1. Parameter change	15. Curvature/spheroidality	29. Inert atmosphere
2. Local quality (+10)	**16. Porous materials (+14)**	30. Phase transitions
3. "The Other Way Around" ("Do It In Reverse")	**17. "Nested Doll" (+17)**	**31. Discarding and recovering (−16)**
4. Mechanics substitution	18. Composite materials	32. Skipping
5. Taking out/separation	**19. Equipotentiality (+18)**	33. Feedback
6. "Intermediary" (+12)	20. Thermal expansion	34. Strong oxidants
7. Segmentation	**21. Color change (−12)**	**35. Cheap, short-living objects (−22)**
8. Preliminary action	22. Flexible shells/thin films	36. "Blessing in Disguise"
9. Another dimension (+10)	**23. Copying (−12)**	37. Anti-weight
10. Asymmetry (+14)	**24. Preliminary anti-action (+15)**	38. Homogeneity
11. Periodic action	**25. Mechanical vibration (−17)**	39. Beforehand cushioning
12. Merging (+21)	**26. Pneumatics and hydraulics (−12)**	40. Continuity of useful action
13. Self-service (+15)	**27. Universality (−7)**	
14. Dynamization	**28. Partial/excessive action (−12)**	

The bolded items were intended to reinforce the inventive principles that had changed the most in their frequency of use.

customer loyalty. The use of this principle has been accelerated through the use of advanced electronic and mechanical assembly techniques.

<u>Self-service</u>—When was the last time someone pumped your gas for you? Unless you live in New Jersey or Oregon (where it is illegal to pump your own), it's been a long time for US residents. Many grocery and department stores also have self-checkout lanes. This degree of self service has been enabled by the use of bar coding

and scanning (informational resources that did not exist a decade ago), and illustrated by the TRIZ lines of evolution to be discussed later.

Preliminary anti-action—In the business world, where unfriendly takeovers have become common, "poison pills" have been written into corporate by laws, making it very expensive for certain companies to be acquired. When we know that something about a system might go wrong, but we are unsure about when, we embed a mechanism to counteract this potentially harmful effect. Automatic kick back counteractions in high-speed saws are one example. Again, our capabilities in this area are now more used because it is possible with advanced sensors and electronics.

Porous materials—Advances in our ability to control porosity has allowed this principle to be more widely applied than it has been historically. Polymer technology in the membrane area now allows us to separate gases and viruses from liquids that would be unheard of in 1950. Our ability to control pore size has also greatly increased our ability to design very specific catalyst systems. This is a case where advances in material science have allowed the more common use of this principle.

Intermediary—This refers to the use of an intermediate material, process, or organization tool to perform a temporary function and then this intermediate being disposed or reused later. A pot holder would be a classic example. In today's business world, we see the increased use of arbitration in management labor disputes. We also see the use of artificial juries to give attorneys a sense of how their cases will be received by real juries to potentially save large sums of money and time in court fees and judgment awards. There are also firms which resell consulting services that provide an artificial barrier between client and consultant. We also see an increased use of transfer molds in manufacturing.

Local Quality—We have discovered that total uniformity is not necessarily the optimum in a product or a process. A surface may need to be uniform, but not necessarily what's underneath—potentially a major cost savings in coatings. A product may not need to be strong everywhere—only at strategic points. Helmets are a good example. Reducing the amount of space and volume required to be strong can save money. Concentrating sales and marketing efforts in certain areas as opposed to widespread geography saves money as well as increases the value received for what is spent. Coupons targeted at specific geographic markets can save money and disrupt competitors at the same time.

Another dimension—As our capabilities to do 3-D modeling has increased, we see increased use of 3-D computer modeling in terms of facility and equipment access design. We also see the use of a third dimension in microchip design and assembly. This is another case where new capabilities (resources) have allowed us to do things we have always wanted to do but could not.

"The other way around"—Many times today, we don't go shopping—the shopping comes to us in the form of web ads. We find repair information on the web instead of a repair person needing to come to our home. Search engine programs that we have designed bring information to us rather than we having to search through much unneeded information.

These principles have decreased in frequency of use the most:

1. Cheap, short living objects
2. Mechanical vibration
3. Discarding and recovering
4. "Blessing in disguise"
5. Partial or excessive action
6. Copying
7. Pneumatics and hydraulics
8. Color changes
9. Beforehand cushioning
10. Dynamism

Let's take a look at this list and try to understand why these principles have decreased in their frequency of use.

Cheap, short living objects—Our interest in minimizing disposal of materials and recycling is a big driver here. The reuse of cloth grocery bags as opposed to paper or plastic is a common every day example. Recycling of aluminum cans is another. In this case, there is also a large economic incentive as the major cost of making aluminum cans is the manufacture of the base metal.

Mechanical vibration—Vibration is just another form of agitation or mixing and doing it in a more advanced (from a TRIZ perspective) way—for example, acoustically or sonically or electronically, yields more uniformity and faster results. We've seen this in any paint store as well as in chemical laboratories. We now know how to achieve many of the same functions with sonic vibration.

Discarding and recovering—As in the case of "cheap, short living objects," our environmental concerns have driven us, when we discard something, to find a way to reuse it. But not discarding in the first place is preferable. Filters that are self-cleaning or that we clean ourselves are another example of not using this principle. The use of this principle in many situations will not disappear as in many cases (such as aluminum cans) the value of the discarded material will make this option attractive.

"Blessing in disguise"—Though this is one of my favorite TRIZ principles, its use has declined as we have implemented process analysis tools such as Six Sigma and Design for Six Sigma, which attempt to eliminate defects in the first place. The elimination of a defect in a process or product is, intuitively, a better approach than making positive use of a negative thing. However, in cases where byproduct negative fields such as heat are produced, finding a way to use this negative byproduct will always be preferred to throwing it away or spending money to dissipate it. The moldy sandwich bag illustrated earlier is an example of the use of this principle combined with that of "copying."

Partial or excessive action—In the past, in many situations where exact control of a system was necessary, we did not have capability to control processes or equipment exactly. In these cases, we used too little to make sure we did not exceed a certain variable or result, or too much to make sure the goal or result was accomplished. We now have far more sophisticated measurement and control systems that minimize the need for using this principle.

Pneumatics and hydraulics—As we will discuss in more detail in the TRIZ lines of evolution section, the use of hydraulics and pneumatics have been slowly replaced with direct electronic controls. The use of electronic actuation is still limited in many situations by the amount of force that can be generated. The ability to measure product and process details more accurately with direct electronic instrumentation and measurement also contributes to the decline in the use of this principle.

Color changes—For decades, the changing of the color of a material (meat, coatings, skin) has been a telling, but indirect, sign of an undesirable phenomena going on. However, in an age where we have greater capability to measure directly the property of concern, the need to measure something indirectly becomes less critical. However, there will always be cases where a color change is an inexpensive way of measuring a phenomenon indirectly.

Beforehand cushioning—In prior years, when we did not have excellent control mechanisms for processes, we often used buffers of one sort or another (excess inventories, contingency plans, pH buffers) to protect against an unusual situation. We now have sophisticated "just in time" processes, more exact control mechanisms and procedures, and faster feedback mechanisms that lessen the need for the use of this principle. However, there will always be situations where we do not understand all the details of how a process or product will respond to a given situation and the incorporation of cushioning of some kind will be necessary and preferable to letting the process or product do whatever it chooses to do.

Dynamism—Though there continue to be many interesting examples of the use of this principle (pricing versus business conditions), its overall ranking has declined. We now require less flexibility in many systems due to our ability to directly control a system electronically (fly by wire) where improvement in measurement technology minimizes the need for dynamic control. With our capabilities to control many business processes, it is less necessary to embed dynamic response into a process.

It is worth emphasizing that all the principles are worth considering, especially in an informal ideation session. They are still the foundation of all the breakthrough patents in the world. Their frequency of use only reflects the current state of business and technology.

Exercises

1. Consider your current product or business. Go through each of the 40 principles and spend some serious time considering how each could improve your process, product, or business.

2. In the next "ideation" or brainstorming session you participate in, make a list of the 40 principles (or use 40 principle stimulation cards) and force the participants to come up with an idea from them.
3. Look in more detail at the ten principles whose use has improved the most over the past 50 years. Are you using them? In what way? Has this benefitted you? In what way? How could you use them more?
4. Think about the ten principles whose use has declined relative to the others. What is your experience? Is it similar? If so, why? If not, why not?

References

1. Hipple J (2005) Chemical engineering progress, April 2005, p 45
2. Hipple J (2005) 40 Principles with examples for chemical engineering. http://www.triz-journal.com/archives/2005/06/06.pdf
3. Domb E. http://www.triz-journal.com/archives/1997/07/b/index.html
4. Mann Darrell (2010) Hands on systematic innovation. Lazarus Press, 215–230

Chapter 10
The TRIZ Contradiction Table

The first form of TRIZ is what we call the "contradiction table," consisting of a listing of a matrix of 39 parameters of physical and engineering systems. From the study of the most inventive patents, it was possible to map contradictions between these parameters and identify the most frequently used inventive principles that were used to resolve these contradictions. This was the basis for identifying the 40 inventive principles discussed in Chap. 9.

A TRIZ contradiction table shows, at the intersection of any these parameters, the most frequently used inventive principles to resolve and deal with that contradiction. In the analysis of millions of patents, we find that there are a limited number of inventive principles that are constantly reused, across virtually all areas of technology and business, to solve the same contradictions that reoccur in all areas.

The way this table is used is relatively straightforward and that is to ask "what property or feature do we want to improve?" (In a physical version of the table, the "y" axis) and when we proceed across the "x" axis to "what property or feature gets worse or deteriorates or gets worse." At the intersection of these parameters will be found the numbers of one of more of the 40 inventive principles most frequently used to resolve that contradiction.

The mechanics of using the table are shown in (Table 10.1).

First we choose the parameter or feature that we want to improve, along the "Y" axis, and then move across to the feature or parameter that gets worse when we try to make this improvement. At that intersection will be the numbers of the inventive principles most frequently used to resolve that contradiction, according to analysis of the global patent literature. These numbers refer to the numbered inventive principles discussed in the last chapter.

Though this original work was focused primarily in the engineering world, it is possible, with a little mind stretching, to convert these parameters into useful characteristics for the non-engineering world. Eventually, a more general algorithm known as ARIZ (Algorithm for Inventive Problem Solving) was developed, and this is beyond the scope of this book. This algorithm is the basis for much of the commercial

J. Hipple, *The Ideal Result: What It Is and How to Achieve It*,
DOI 10.1007/978-1-4614-3707-9_10, © Springer Science+Business Media New York 2012

Table 10.1 Mechanics of using the TRIZ contradiction table

| Improving feature ↓ | Worsening feature → | | | |
	Parameter 1	Parameter 2	Parameter 3	Parameter 4
Parameter 1	xxxxx	3,19, 35, 40	8,15, 17, 35	12, 15, 17, 28
Parameter 2	2, 3, 35, 40	xxxxx	4, 17, 30, 35	9, 17, 31, 35
Parameter 3	4, 15, 17, 31	1, 2, 15, 17, 30	xxxxx	1, 15, 17, 24
Parameter 4	8, 30, 31, 35	2, 31, 35, 40	1, 3, 4, 17, 19	xxxxx

TRIZ software on the market. My experience with organizational problem solving is that this degree of sophistication is seldom needed.

These original table design parameters and a definition of what they mean are provided here. These definitions include work by TRIZ colleagues, Ellen Domb and Darrell Mann [2, 3]. Many of these definitions will seem trivial for most engineers and scientists but they are useful for others in understanding the parameter and possibly being able to generalize it in a nontechnical sense. These are the parameters used to characterize the improving and worsening feature of an object, system, or service.

The TRIZ contradiction table in this book is the most up-to-date public version of the table and is based on the work of Mann [1] to update the patent literature analysis to the present day. This table uses a grouping of parameters and has many of them in a different *order* than the original table, whose content was based on the analysis of only a few hundred thousand patents through the early 1950s. In discussing the use of the contradiction table with someone who may be using the older table, it is best to describe the parameters you are using by name, not by number, so that an "apples and apples" conversation occurs. This newer table includes nine additional parameters which were not in the original table.

TRIZ Parameter Definitions

These numbers refer to those on the far left and at the top of either table. The list is in the same order, both vertically and across.

1. Weight of a moving object: Mass of an object in a gravitational field. The force that the body exerts on its support or suspension.
2. Weight of a stationary object: Same plus the surface on which it rests
3. Length of a moving object: Any one linear dimension, not necessarily the longest
4. Length of stationary object: Same
5. Area of moving object: A geometric characteristic described by the part of a plane enclosed by a line. The part of a surface occupied by the object. Square measure of the surface, either internal or external, of an object
6. Area of a stationary object: Same
7. Volume of moving object: Cubic measure of space occupied by the object. Length × width × height for a rectangular object. Height × area for cylinder.

8. <u>Volume of stationary object</u>: Same
9. <u>Shape</u>: External contours; appearance of a system
10. <u>Quantity of substance or matter</u>: Number or amount of a system's materials, substances, parts, or subsystems which might be changed fully or partially, permanently or temporarily.
11. <u>Amount of information</u>
12. <u>Duration of action by a moving object</u>: Time that an object or system can perform a function. Service life. Mean time between failures. Durability.
13. <u>Duration of action by a stationary object</u>: Same
14. <u>Speed</u>: Velocity of an object. Rate of a process or action in time.
15. <u>Force/torque</u>: Interaction between systems. Mass×acceleration. Any interaction intended to change an object's position
16. <u>Use of energy by a moving object</u>: Measure of an object or system's capacity for work. Classically defines as force×distance. Includes energy provided by the supersystem in which the object or system is a part of. Energy required doing a particular job.
17. <u>Use of energy by a stationary object</u>: Same
18. <u>Power</u>: Time rate at which work is performed
19. <u>Stress or pressure</u>: Force per unit area; tension
20. <u>Strength</u>: Extent to which an object or system is able to resist changing in response to force. Resistance to breakage.
21. <u>Stability of an object's composition</u>: Wholeness or integrity of the system; relationship of the system's elements. Wear, chemical decomposition, disassembly are all examples. Increase in entropy. Organizational stability.
22. <u>Temperature</u>: Thermal condition of an object or system. Can include other parameters such as heat capacity, thermal conductivity, or any other variable that affects temperature.
23. <u>Illumination intensity</u>: Light flux per unit area, brightness, light quality, etc.
24. <u>Function efficiency</u>: The degree to which a function is performed vs. its maximum potential. Number of energy conversions used in a system.
25. <u>Loss of substance</u>: Partial or complete, permanent or temporary, loss of some of an object or system's materials, substances, parts, or sub-systems
26. <u>Loss of time</u>: Time is the duration of an activity. Improving loss of time is the reduction of time used for an activity. "Cycle reduction" time.
27. <u>Loss of energy</u>: Use of energy that does not contribute to the job being done
28. <u>Loss of information</u>: Partial or complete, permanent or temporary, loss of data in or by a system. Includes sensory data such as aroma, touch, and feel.
29. <u>Noise</u>: Level of acoustical emissions from an operation, piece of equipment, or people. Could be at several different frequencies.
30. <u>Harmful Emissions</u>: Undesirable material emissions from a process or product in use.
31. <u>Harmful effects generated by the system</u>: Harmful effects (reducing efficiency or quality) generated by the object or system as part of its operation.
32. <u>Adaptability/versatility</u>: Extent to which an object or system positively responds to external changes. System that can be used in multiple ways or for different functions, under a variety of circumstances.

33. Compatibility/Connectability: Ability of an object or system to interact with other parts of a system or interacting objects without negative side effects. The ability of a physical component to easily connect with other components in its system or supersystem.

34. Ease of operation/trainability/controllability: Simplicity, high yield, ease of doing "right"

35. Reliability/robustness: A system's ability to perform its intended function in predictable ways and conditions.

36. Repairability: Quality characteristics such as convenience, comfort, simplicity, time to repair faults, failures, defects in a system

37. Security: The ability of an object or system to be secure from negative, saboteurial actions. Ability of a software or control system to resist outside interference. An organization's capability to resist business threats.

38. Safety/vulnerability: Ability of a system or process to perform its function without harming people or other systems. Ability of a system or organization to repel external threats.

39. Aesthetics/appearance: The "beauty" or "pleasing functionality" of a system or object as perceived by users and others.

40. Harmful effects acting on the system or process: Susceptibility of a system to externally generated harmful effects

41. Manufacturability: Degree of facility, comfort, or effortlessness in manufacturing or fabricating the object or system.

42. Manufacturing precision: Extent to which the actual characteristics of an object or system match the specified or required characteristics.

43. Automation: Extent to which a system or object performs its function without human assistance.

44. Productivity: Number of functions or operations performed per unit time. Time required for a unit operation or function. Output per unit time.

45. System complexity: Number of elements and element inter-relationships within a system. User is included in this analysis. Difficulty of mastering a system is a measure of its complexity.

46. Control complexity: Number of interfaces required to control a process or system. How complicated a control system must be to respond to upsets or external effects.

47. Ability to detect or measure: Measurement of the parameters of a system. Measurement of components that interfere with each other.

48. Measurement accuracy: Closeness of measured value to actual value of a property or system.

This list includes nine additional parameters (compared to the original table) that simply were not of any significance in the 1950s and would not have shown up to any significant degree in a patent search 50 years ago. These include safety and vulnerability, security, control complexity, measurement precision, aesthetics and appearance, repairabilty, function efficiency, noise, and harmful emissions.

Remember, these parameters are the defining aspects of a system or product. In difficult problems, improvement in the value of one of these parameters is desired.

When we try doing this, a different parameter gets worse. As a reminder, the desired improving parameter is found on the "Y" (vertical) axis of the table, and the worsening parameter is identified along the top (horizontal) or "X" axis. At the intersection, as shown in Fig. 9.2, are the inventive principles (reviewed in Chap. 8) most often used to resolve that contradiction.

The contradiction table is shown on pp. 114–131. The traditional table, with only 39 parameters, can be found in numerous texts and resources. See references [3, 4, 5, 6] in the appendix.

Using the TRIZ Contradiction Table

Let's look in more detail how we use this table. As mentioned previously, we identify the parameter of the system, product, or process that we want to improve (on the "y" vertical axis) and then move horizontally ("x" axis) across the table to the parameter or feature that degrades or gets worse when we try to make this improvement. At the intersection of these two parameters is a list of 4–5 principles which are the most frequently used inventive principles to resolve that particular contradiction, based on the analysis of the patent literature.

Most of the time, each of the 4–5 suggested principles will lead to at least one new breakthrough idea each. Having a diverse group of people, including people who are not directly involved with the problem, will increase the effectiveness of the use of the suggested principles. It is a useful tactic to have individuals make a sketch of the problem solution suggested by a given principle and then share it with the rest of the group, which will then build on the suggestion. This is where TRIZ has some overlap with other inventive processes. Making picture cards that illustrate the inventive principles is another technique. If it is not possible to link one of the suggested inventive principles with an idea, just move on.

Let's look at a simple example of the use of the table. As Boeing was in the process of developing a stretched version of its 737 aircraft, the weight, passenger, and fuel load of the aircraft steadily increased, requiring a larger engine. By normal thought processes, this obviously requires a larger engine cross-sectional area if we define area by the normal geometric equation, $A = \Pi r^2$. As this aircraft stretched from its original version to a 737–200 and subsequently a 737–300, the next leap to a 737–400 required an engine cowling frame that violated a Federal Aviation Administration (FAA) requirement for a certain level of ground clearance between the engine cowling and the ground. This is due to concern about an uneven landing and the possibility in rough weather that the engine frame would impact the ground. This was a significant challenge for Boeing engineers and was a bottleneck in the 737–400 launch.

Let's take a look at this problem with the contradiction table (in this particular case, it makes no difference between the use of the traditional versus the newer table since the parameter numbers are the same). What are we trying to improve? Remember that the parameters in the table are designed to be general and applicable to many different systems, products, and processes. I think we would agree that

Fig. 10.1 Asymmetric engine cowling

"area of a moving object" is the appropriate parameter, which is #5 in either table. What parameter is getting worse? This would be "diameter of a moving object", #3 across the top/horizontal axis. In other words, we want a larger area without a larger diameter. At this intersection point in the table, we find listed the following inventive principles:

#1: Segmentation
#4: Asymmetry
#14: Curvature
#15: Dynamization
#18: Mechanical vibration

The next time you fly on a Boeing 737 aircraft (there are very few 737–100, 200, and 300's left in service), take a look out the window and you will see how Boeing solved this problem, using principle #4: asymmetry (Fig. 10.1).

It would be possible to solve this problem with all but one (mechanical vibration?) of the other principles as well. For example, we could design the cowling with a dynamic, segmented metal frame which could be changed when the aircraft is landing. In a high level sense, curvature was also used in the solution.

Asymmetry is an interesting inventive principle and, as noted earlier, is one of the inventive principles whose use has increased the most in 50 years. Why is it a powerful inventive principle? Because it breaks the natural symmetrical thinking patterns that we have in our brain. It is easy to envision circles, squares, diamonds, triangles, and rectangles. They are easy to draw and reproduce. They are *symmetrical*. Our recent emphasis on reproducibility on statistical process control and minimizing variation reinforces this natural tendency.

Since the contradiction table is showing only the top 4–5 principles most often suggested by the trends in the patent literature, there is always the possibility that other principles may provide useful ideas. Assuming time is not a constraint, it is suggested that all the principles be reviewed. In a group setting, individuals or small teams can be assigned one or more of the principles and ask them to suggest a possible answer to the contradiction.

Though the TRIZ contradiction table is a simple and intellectually attractive tool, it is one that can be misused and lead to quickly conceived, but potentially incorrect or incompletely thought out ideas. For example, we often have multiple contradictions within a system and insufficient time is spent discussing what the "limiting" contradiction might be, or whether we are actually working on the correct contradiction or problem. A system that is complex with multiple contradictions and many functional relationships is best analyzed and handled with one of the more sophisticated TRIZ algorithms (see Chap. 15). It is also easy to jump to a conclusion about an "answer" (the inventive principles to consider in solving the problem) without thinking in depth about whether the contradiction has been adequately or properly described. For example, let's think about the basic contradiction of car weight versus comfort. Most people would agree with the perception that a "heavier" or bigger car is more comfortable but would also have lower fuel economy. If that's all we thought about, how would we use the table? We could consider the parameter we want to improve to be #1, "weight of a moving object." Fuel economy could be matched with parameter #16, "use of energy by moving object." At the intersection of these two parameters in the table, we would find the following suggested inventive principles:

#35: Parameter change
#10: Preliminary action
#19: Periodic action
#3: Local quality

If we look at what both Honda and General Motors have recently done in their more expensive, heavier cars, they have introduced an engine which uses only the number of cylinders necessary to deliver the horsepower needed at that time, which is principle #19, periodic action.

How would you use the other principles in a proposed concept of solution? Don't worry about cost, etc. Just think about a general concept of a possible solution.

#3: Local quality_____
#10: Preliminary action_____
#35: Parameter change_____

Suppose we chose the improving factor, in plain normal English, to be comfort. Now this is a term that is not easy to turn into an engineering parameter and makes the use of such a table much more challenging, in part because not everyone will define comfort in the same way. We would need to think further about what actually causes discomfort in a small car. Is it elbow room? Is it bouncing up and down? Is it lack of noise from the outside? Some more thinking and analysis is required to use the table here. If we were to decide that "elbow room" is what we're concerned about, we might revisit the contradiction parameter and look at "volume of a moving (or stationary) object and see where that might lead.

When a system is complex, it is usually more productive and efficient to use one of the system modeling tools now developed for TRIZ (Su-Field modeling, cause and effect modeling), which will be briefly discussed in Chap. 15. It is also not unusual to have a complex system or process which has numerous contradictions, as well as multiple functional relationships, and deciding which ones to focus on can be a challenge. These tools allow diagramming of a complex system and give a clearer indication of where the key issues are and which TRIZ tool should be applied. However, there is a middle ground that I have found effective in many situations and that is to identify clearly several sets of contradictions that may characterize a system and map each set using the contradiction table. This usually produces an overlap where a few of the inventive principles tend to appear multiple times and provides a focus for problem solving.

As a simple example, let's go back to the automobile design issues discussed previously. Using the table, we could look at the contradiction of improving feature "noise" (parameter #29) vs. "weight of a moving object" (parameter #1) and see the following suggested principles:

1. Porous materials (principle #31)
2. Preliminary anti-action (principle #9)
3. Local quality (principle #3)
4. "Blessing in Disguise" (principle #22)
5. "Other Way Around" (principle #13)

If we looked at this contradiction as improving feature "reducing weight" (parameter #1) versus "noise" (parameter #29), we would see the following suggested principles:

1. Parameter change (principle #35)
2. Taking out (trimming) (principle # 2)
3. Self-service (principle #25)
4. "Other Way Around" (principle #13)

Now it is possible to think of various ways of dealing with this contradiction from any of these principles, but the fact that principle #13 shows up in both ways of looking at the problem might suggest that we start there. In fact the noise cancelling head phones now on the market use this exact principle. They measure the noise and instantly output an exactly 180° signal that cancels out the noise. The question then is how this could be done in the confines of car as opposed to many pounds of insulation.

We have now added another box for achieving the Ideal Result (Fig. 10.2).

Fig. 10.2 TRIZ problem solving with ideal result, resources, trimming, and contradiction resolution

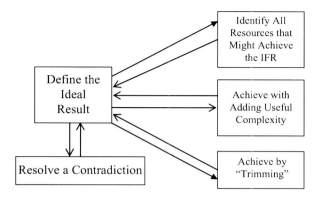

Exercises

1. The next time you are involved in a "brainstorming" session, bring a list of the TRIZ 40 Inventive Principles into the room with you and randomly suggest ideas based on them or ask the group how they would apply that principle (you may have to explain the basics of the principle).
2. Whenever you are involved in a product or process design issue, make sure you have defined it in terms of a contradiction, not just making something "better". Focus the meeting in terms of what the contradiction(s) is or are. Don't give up until this is accomplished. Remind everyone that breakthrough patents and inventions come only from resolving contradictions. Try to define the contradiction in terms of physical aspects of the system.
3. Try out the contradiction table on some simple, well defined contradictions. Use it to reverse engineer some recent breakthrough that your group has accomplished and see if the solution you came up with is generally suggested by the inventive principles. Then evaluate how much time might have been saved had you used this simple tool.
4. If your system has multiple contradictions, try mapping the contradiction several different ways via the table and see if there are a few principles that show up no matter how you look at the problem. Apply these to come up with a solution.

References

1. Mann D (2004) Comparing the classical and new contradiction matrix. TRIZ Journal
2. Domb E. The 39 features of Altshuller's contradiction matrix. http://www.triz.journal.com/archives/1998/11/index.htm
3. Mann D. http://www.triz-journal.com/archives/2004/07/05.pdf

Improving Feature			Worsening Feature →	**Physical**													
				Weight of Moving object	Weight of Stationary Object	Length/Angle of Moving Object	Length/Angle of Stationary Object	Area of Moving Object	Area of Stationary Object	Volume of Moving Object	Volume of Stationary Object	Shape	Amount of Substance	Amount of Information	Duration of Action of Moving Object	Duration of Action of Stationary Object	Speed
				1	2	3	4	5	6	7	8	9	10	11	12	13	14
Physical	1	Weight of Moving Object			3 19 35 40 1 26 2	17 15 8 35 34 28 29 30 40	15 17 28 12 35 29 30	28 17 29 35 1 31 4	17 28 1 29 35 15 31 4	28 29 7 40 35 31 2	40 35 2 4 7	3 35 14 17 7	31 28 26 7 2 3 5 40	2 5 7 4 34 10	10 5 34 16 2	10 5 28 35 16 2	15 2 25 19 38 18
	2	Weight of Stationary Object		35 3 40 2 31 1 26		17 4 30 35 3 5	17 35 9 31 13 3 5	17 3 30 7 35 4 14	17 14 3 35 30 4 9 40 13	14 13 3 40 35 5 30	31 35 7 23 1 30	13 7 9 30 31 29 10	35 31 5 18 25 2	28 13 7 26 2 17	3 35 10 12 4 17 14	40 35 31 6 19 27 2	3 35 17 3 36 2
	3	Length/Angle of Moving Object		31 4 17 15 34 8 29 30 1	1 2 17 15 30 4 5		1 17 15 24 13 30	1 3 29 30 35	17 3 15 1 29	17 4 17 35 1 31 30 2	17 31 3 19 14 4 30	10 17 14 12 30 31	14 13 4 30 31	19 17 10 1 17 13 15	10 35 1 3 9 2	14 1 13 4 17 12 4 3
	4	Length/Angle of Stationary Object		35 30 31 8 28 29 40 1	35 31 40 2 28 29 4 3	3 1 4 19 17 35		3 4 19 17 35	17 40 35 10 31 14 4 7	35 30 14 7 15 17	14 35 17 2 4 7 3	13 14 15 7 17 3 30	4 3 31 25 17 14	7 17 2 22 28 13	35 3 29 2 31 7 19	35 10 1 3 2 25 4 5	3 14 4 13 18 31 9
	5	Area of Moving Object		31 17 3 4 1 18 40 14 30	17 15 3 31 2 4 29 1	14 15 4 18 1 17 30 13	14 17 15 4 13		17 1 4 3 24 5 2	14 13 3 18 1 31 3 18	14 17 7 3 31 3 1 18	35 4 14 17 15 34 29 1	31 30 3 16 29 1 5 19	17 15 14 32 1 3	3 19 18 16 5 2	1 3 19 2 6 5	14 3 34 29 28 30 13
	6	Area of Stationary Object		14 31 17 19 4 13 3 12	35 14 31 30 17 4 18	17 19 3 13 14	17 14 4 7 9 24 13 26	4 31 7 19 15 14 3 13		17 18 14 7 30 13 26	14 28 26 13 4 35 17	17 5 4 7 28 26 14	35 16 17 21 18 40	30 35 7 24	13 19 37 10 35 14 5	30 30 20 25 13 37 26 5	26 28 17 13 14 5 2 33 3
	7	Volume of Moving Object		31 35 40 2 30 29 26 19	31 40 35 26 2 13 30	1 7 4 35 3 29 15 13 30	7 15 4 3 1 35 19 10	17 4 7 1 31 35 19 10	17 14 4 3 31 24 36 35	7 35	35 14 28 23 24 13	15 1 4 19 29 14 38 25	30 31 7 29 36	10 2 28 7 32 3 15 26	4 36 35 31 6 1 30 28 5	30 31 1 35 4 28 38 5	29 4 28 1 35 38 3 13 14
	8	Volume of Stationary Object		31 30 40 35 3 2 4 19	35 40 31 9 14 13 3 4 26	14 30 15 3 4 35 2 19	35 2 30 4 14 8 19 28 26	15 14 4 30 13 3 7 28	15 14 35 34 13 15 26 17	14 4 7 30 13 1 26	14 35 3 28 2 30 7		7 35 2 30 13	31 33 31 40 5 13 17	31 3 31 8 26 2	35 19 1 38 15 34 28 3 1 34	35 40 2 38 28
	9	Shape		29 30 3 10 40 8 31 35	15 3 10 31 26 35 40	4 14 29 5 15 13 2 7	17 14 4 13 5 7 31	4 17 5 2 14 7 31 32	17 14 5 28 2 32 4	14 4 15 3 7 32	14 4 7 1 2 35 5 32		3 31 30 36 5 4 22	17 7 3 32 24 1	14 28 25 30 26 31 9	3 30 28 35 13 5 22	15 35 10 3 18 4 1
	10	Amount of Substance		35 40 6 18 9 2 31 18	35 40 18 5 2 8	29 3 17 35 14 2 18 36 31	35 31 3 17 14 2 40	15 14 17 31 35 4 30 29 18	17 31 4 18 35 4 30 29	3 2 38 25 38 24	35 7 14 3 31		17 37 4 31 13	35 40 34 10 3 7 18 19 12 2	34 3 38
	11	Amount of Information		28 17 13 7 1 35 2	28 26 35 3 2	7 32 13 17 2 3 14	7 32 17 3 2 14	7 17 32 2 24 3 28	32 2 3 24 7 28	7 19 26 3 32 24 28 2	26 32 3 2 24 28	7 17 3 32 28 13	7 17 32 13 28 35		7 3 32 13 12 24 19	7 17 32 13 13 24	37 3 31 28 12 5
Performance	12	Duration of Action of Moving Object		15 19 5 8 31 34 35	35 3 31 34 8 4 2	17 8 19 9 35 2 12 24	3 17 12 9 35 2 19 13	19 17 8 9 24 13	3 17 9 24 12 19 13	19 10 30 7 14 2 13	10 30 35 12 13 4 2 3	17 14 28 10 1 26 25	3 40 17 35 6 10 13	7 2 32 3 24 10 25		10 2 24 20 13 4 17 6	3 35 5 13 17 4 37 9
	13	Duration of Action of Stationary Object		35 31 8 19 4 15 34	6 2 31 19 3 34	17 40 19 2 35 3 8	40 35 1 9 17 2 13	35 18 19 14 2	35 31 30 7 14	35 19 18 3 13 17 4	35 40 31 3 34 38 19 13	17 3 40 14 33 13 10 7	35 31 3 40 17 13 6	35 24 28 4 25 29 34		35 28 29 3 4 14 13
	14	Speed		13 14 8 28 1 17 2 38	1 13 2 10 35 3	13 17 28 2 29 14 1	17 15 30 2 14 1	17 15 4 30 3 5 34	14 5 17 1 4 13	28 2 7 34 5 14 4	28 5 2 35 7 9 1 18	17 7 15 18 35 3 4 2	2 35 19 5 10 38 9	7 2 10 5 37 28 3	35 40 19 3 5 13	3 13 35 5 2 24	
	15	Force/Torque		8 1 9 13 37 28 31 35 18	13 28 1 35 40 18	9 13 28 1 35 19 28 36 29	35 28 17 9 19 28 36 29	15 17 10 14 19 3 29 39 40	1 3 17 40 37 18 9 35	12 15 9 35 37 14 4	18 37 35 3 2 10 36	18 1 35 36 31 8 36	13 17 37 3 1	19 10 2 12 28	2 10 13 3 12 19 26	13 15 9 28 12 2
	16	Energy Used by Moving Object		28 35 19 12 31 18 5	28 35 5 12 18 31 13	28 12 15 35 17 19	35 2 19 15 28	15 19 4 3 25 14	17 24 17 15 3	35 13 28 35 35 7 2	29 2 3 12 15 28	19 28 35 6 18 16 38 30	15 12 10 37	18 28 35 6 15 12 10	35 28 13 19 12 10 19	35 28 13 19 5 8
	17	Energy Used by Stationary Object		35 28 13 8 3 19	19 13 35 9 28 6	17 4 12 3 24 14	4 17 9 19 3 14	3 4 13 5 12 24	4 17 3 14 16 19	2 35 13 19 13 18 28 4	35 39 19 2 13 5 10 36	7 35 24 30 13 5 10 36	35 31 3 24 28 13 4 9	2 19 17 20 7	40 35 3 19 4 28 24	40 35 3 17 28 26 24	1 3 28 19 13 25
	18	Power		8 38 2 25 31 19 28 35	19 2 35 31 26 28 29 17 27	1 17 35 10 37 36 28 30 29	17 14 1 35 4 10 28 29 30	19 38 35 2 25 3 15 4	17 19 38 13 3 15 25 32 2	19 38 2 35 3 15 25 6 15 4 14	19 35 25 36 6 30 15 3 14	29 14 15 1 35 40 4 3 36	15 39 38 4 3 40 18 28 24	10 28 19 12 24 37	19 35 10 38 4 28 1 21	38 35 10 4 28 19 16	15 2 19 35 3 14 24 1 13
	19	Stress/Pressure		40 35 31 10 36 37 2 17	35 10 13 31 29 40 2 17	35 9 40 17 3 14 4 13	3 14 17 35 40 4 9 1	14 10 34 35 11 28 15 3	40 14 35 10 37 17 3	35 10 40 34 15 14 4 17	35 4 40 3 30 24 14	35 4 40 3 14 31 15	35 34 35 9 2 14 13 12 5	28 2 26 7 24	19 3 35 4 13 14 2	13 3 14 12 5 12	13 6 14 29 36
	20	Strength		40 31 17 8 1 35 3 4	40 31 2 1 17 26 35 3	17 35 40 1 4 15 8	17 35 9 37 14 4 40 15	14 17 3 7 19 4 40 5	14 17 9 40 4 4 5	4 7 17 14 35 4 13	14 9 17 4 31 10 15 31	40 4 9 35 7 17 30 25	30 17 31 9 13 29 3	17 2 32 3 28 26	35 40 3 26 17 4 13	35 3 5 24 26 4 13 40	14 28 8 13 12 26 2
	21	Stability		40 35 31 5 2 39 17 24 8	40 35 31 17 39 1 24 4	1 35 13 15 28 4 25	17 4 35 37 13 1 40	35 4 37 4 3 2 12	17 3 14 13 17	24 5 39 35 10 19 28 25	25 4 5 14 5	1 4 35 17 3 18 21	5 24 31 40 35 15 39 13	2 7 10 25 5	10 13 5 34 19 7 40	10 40 3 39 23 6 13 7	40 28 25 13 24 10 33 15 18
	22	Temperature		36 31 6 35 38 30 22 19 40	31 35 3 32 36 22 40 4	15 19 9 3 33 43 4 1	19 3 15 31 9 35 4 1	3 35 40 19 18 1 39 31	35 40 1 18 39 34 17	35 40 3 14 6 30	35 40 31 3 4 3 22 31	14 19 32 39 39 17 15 19	30 31 3 35 22 31	5 37 7 10 26 19 32	19 15 13 39 1 18 30 9 3	19 36 40 3 9 1 13 25 28	28 14 36 2 14 13
	23	Illumination Intensity		19 1 24 32 31 39 35	35 32 2 19 31 1 5 30	14 19 32 35 17 24 1	14 17 2 35 24 19 1	14 17 24 35 19 32 26 4 1	14 1 4 35 24 32 39 1 26	14 24 13 10 19 2 32	14 24 13 10 19 2 32	3 30 13 24 32 35 5 14 1	1 19 35 14 24 28 3	2 25 19 32 6 3	19 2 6 35 28 4 25	2 6 10 35 28 4	19 10 13 28 35 4 5

40 Inventive Principles

#	Principle	#	Principle
1	Segmentation	21	Skipping
2	Taking Out/Separation	22	'Blessing In Disguise'
3	Local Quality	23	Feedback
4	Asymmetry	24	Intermediary
5	Merging	25	Self-Service
6	Universality	26	Copying
7	'Nested Doll'	27	Cheap Short-Living Objects
8	Anti-Weight	28	Mechanics Substitution
9	Preliminary Anti-Action	29	Pneumatics And Hydraulics
10	Preliminary Action	30	Flexible Shells And Thin Films
11	Beforehand Cushioning	31	Porous Materials
12	Equipotentiality	32	Colour Changes
13	'The Other Way Around'	33	Homogeneity
14	Curvature	34	Discarding And Recovering
15	Dynamization	35	Parameter Changes
16	Partial Or Excessive Actions	36	Phase Transitions
17	Another Dimension	37	Thermal Expansion
18	Mechanical Vibration	38	Strong Oxidants
19	Periodic Action	39	Inert Atmosphere
20	Continuity Of Useful Action	40	Composite Materials

Performance (Contradiction Matrix — parameters 15–23)

Column parameters:
15 Force/Torque · 16 Energy Used by Moving Object · 17 Energy Used by Stationary Object · 18 Power · 19 Stress/Pressure · 20 Strength · 21 Stability · 22 Temperature · 23 Illumination Intensity

θ	15	16	17	18	19	20	21	22	23
1	10 30 35 28 8 37 18	36 10 19 3 34 31 12	35 15 1 28 7 14 34 39	19 30 36 31 12 18 14 25 1	10 40 30 36 37 31 4 3 12 5	28 31 40 35 10 30 18 14 4	35 1 30 39 29 21 7 19 12	2 40 6 31 36 29 4 38 32	1 35 32 38 13 19 23
2	35 9 8 3 40 13	3 17 14 19 18 30	12 13 1 19 35	13 35 19 9 35	35 8 3 13 40 37 31 4 2 5	35 31 8 40 4	15 5 17 30 4 1 12	35 3 36 19 32 30 25	19 35 26 24 32 30 25 31 7
3	17 4 14 10 7 12 2	1 35 17 14 3 25 8 24 22	12 13 1 24 15	1 35 14 3 29 19 4	35 8 3 14 8 17 29	35 31 8 40 17 4 29 13 5	1 15 3 14 17 34 12 35 40	10 15 35 3 36	19 1 32 5 24
4	10 17 35 9 28 4 12 2	3 17 19 28 4 28	35 3 30 31 1 38	17 19 35 12 13 8 3 31	35 17 3 14 3 1	14 40 35 15 30	2 11 13 35 39 24 40 26	35 15 10 24 32 3 15	19 13 15 32 30 1 35
5	3 2 35 19 17 1	7 10 3 22 6 24	3 17 19 36 5 1 38	19 40 18 1 30 10 32	15 30 10 40 17 3 40 37 36	3 15 40 14 35 40 14 8 36	2 11 13 39 24 40 26	3 15 19 32 13 36 4	19 13 15 30 1 35
6	14 1 35 18 17 36	7 10 3 2 6 24	3 17 19 5 1 38	19 40 18 1 30 10 32	15 30 10 4 3 40 37	40 2 17 39 14 8 36	28 10 1 39 14 8 2	9 18 3 5 34 1 23 16	40 24 35 17 3 32
7	15 1 35 3 36 4 14 37 19	35 19 13 38 33 25 10	35 40 3 1 2 28	35 19 6 10 13 24 28 31	35 9 3 40 1 3 40 37 36	14 5 7 4 31 30 40	28 10 1 39 19 35 30	10 39 18 31 34 1 2 3 16	35 2 28 25 17 3 32
8	37 9 18 12 2 29 5	35 19 2 40 31 13	35 2 40 13 12	35 19 6 10 40 24	35 28 34 4 2 12	13 14 19 17 40 5 12	40 35 31 34 1 2 30 5 28	3 26 4 35 6 19	35 24 28 5 7 38
9	14 17 35 9 2 3 12 13 5	3 14 28 24 6 34	35 14 15 1 3 7 24 1	4 6 2 30 13 14 7	3 14 9 15 2 4 35	35 7 40 14 4 3 2 1	4 37 3 10 2 13 25	31 19 2 32 24 15 1	35 15 28 32 13 24 1
10	36 14 40 3 16 13 2 18	7 34 35 3 13 18	35 14 28 31	7 10 3 2 6 24 12	40 9 3 35 17 3 13 4 36	14 17 7 40 3 12 1	24 37 2 3 1 25	2 13 10 19 17	35 28 30 3 17
11	26 17 2 1 28	7 10 3 2 6 24	10 6 3 2 24	7 10 3 26 24 12	2 17 7 24 12 1 26	2 17 72 1 26 24	24 37 3 10 13 25	2 13 10 19 17	15 19 4 32
12	19 2 16 13 15 12 9	7 10 3 6 24	35 6 18 28 19 12	7 10 3 28 13 12 38 13 12	12 19 40 3 1 26	35 3 17 14 12 27 4 19	35 24 40 13 33 12 19	19 35 13 3 36 17 39	35 19 2 4
13	17 40 35 9 7 5	13 25 40 24 3 27 35	35 40 13 2 19 7 2 30	35 2 13 16 40	7 4 40 12 14 13	35 9 3 9 12 4 19 27	35 5 39 3 12 24 31	19 24 35 40 36 15 16 2 30	35 40 2 19 7 13
14	19 13 15 29 3 18 15 17	28 28 30 28 12 15 38	35 19 13 38	19 35 3 37 1 28 18	21 18 40 38 18 12 5	35 14 3 19 27	35 3 39 12 18 1	19 5 3 2 14 32 13 30	19 35 35 28 28 5 23 7 31 3 16 23
15	21 2 19 35 26 16 18 24 3	2 19 3 13 5	15 28 13 2 35	19 35 39 8 37 1 28 18	18 3 40 1 28 18	27 3 10 3 18 40 4	35 4 40 17 14 18 1	24 19 36 2 3 31 96	35 19 24 35 2 19
16	9 36 37 35 5 17	2 19 3 13 5	15 28 13 2 35	28 1 37	28 37 7 3	17 1 40 33	35 5 10 34 31 10 2 33	24 12 25 19 2	35 24 14 3 1
17	2 19 15 35 36 1 13 13 14	19 6 37 35 16 37 1	19 15 3 26 1	35 29 10 14 28 4	35 10 3 21 35 26 13	30 14 10 40 3 19 35 38	19 13 39 24 17	32 35 19 14 22	19 35 28 26 21 7 31 3 16 23
18	35 14 9 12 4 36	17 11 24 12 37 29 35	17 14 4 32 24 37	35 29 10 14 28	3 17 40 9 40 14	17 9 18 36 28 35	35 19 2 31 32	35 36 21 10 31 96	24 12 21 35 19 2
19	40 9 35 25 27 4 13	5 17 10 19 18	35 17 14 17 13	40 5 4 10 29 26	35 40 24 17 25 28	35 4 40 31 17 24 31 5	35 5 19 2 30 31 96	35 40 9 31 96	35 24 28 24 2 19
20	24 21 10 1 35 17 24 10	13 35 19 18 9 24	13 1 32 32	40 14 28 26	40 3 35 31	2 9 5 35 40 31 49 3	28 32 31 3 27 40	35 40 3 1 24 18	35 19 32 32 4 13 1 40 28 4 5

Improving Feature / Worsening Feature		24 Function Efficiency	25 Loss of Substance	26 Loss of Time	27 Loss of Energy	28 Loss of Information	29 Noise	30 Harmful Emissions	31 Other Harmful Effects Generated by System	32 Adaptability/Versatility	33 Compatibility/Connectability	34 Transability/Operability/Controllability
Physical		**Efficiency**									**Performance**	
1 Weight of Moving Object		1 2 28 3 35 25	5 35 31 24 3 4 7 40	10 19 28 20 35 1 2 3 16	19 6 18 16 13 34 12	10 24 11 35 25	35 2 25 13 5 39 14	2 30 39 25 10 21 31 35 27	2 35 10 39 3 31 22 24 27	29 28 15 2	10 15 13 5 24 25 26	2 15 3 11 25 23 35 24 1
2 Weight of Stationary Object		35 31 7 3 13 40	13 14 5 3 17 12	10 35 19 25 28 26 20	28 35 7 31 13 32 39	10 15 35 3 32 7 26	14 35 31 1 9	40 33 39 14 31 24	40 3 39 2 31 28 8 15 39	35 3 15 19 17 12	2 13 24 3 17 39	13 1 32 14 4 26 6
3 Length/Angle of Moving Object		17 35 19 3 4 1 13 28	1 4 35 29 28 3 10 23	15 19 2 10 29 4 3	35 14 1 7 3 39 19	1 24 25 2 3 32	17 3 13 1 19 28 35	10 35 38 7 19 24 30	3 17 15 2 19 25 33 39	15 14 17 16 13 3 19	23 28 17 3 10 1	15 29 35 10 1 4 7 13
4 Length/Angle of Stationary Object		3 35 28 30 15 24 09	28 24 35 12 10 3 4 17	29 10 17 37 18	1 17 28 3 3 39 19	26 24 13 3 26 14 15 17	14 1 3 35 25 11 9	1 3 21 18 13 14 2 4	35 17 1 15 17 31 24 14	1 19 35 15 17 31 24	1 7 3 31 35	10 25 26 2 12 37
5 Area of Moving Object		3 17 1 15 19 3 24	2 17 35 3 34	2 26 35 39 4 24	17 15 30 2 19	3 2 17 24 26 30 13 14	3 35 1 14 25 31 9	2 19 31 25 18 34 35 38	17 2 18 15 39 31 3	15 3 35 1 30 40 2 25	28 3 17 10 23 15 4 2	15 17 25 13 1 16 19
6 Area of Stationary Object		28 3 13 12 1 14 4 15 19	17 14 12 10 18 39 13 34	10 18 3 25 17 18 3 25	7 28 38 7 28 40 35	28 3 2 10 32 3 24	23 17 31 27 31	35 31 17 1 7 18	49 17 1 15 12 9 18 28	22 35 25 37 3	24 2 17 5 15 35	10 4 24 25 26 16
7 Volume of Moving Object		28 10 13 25 34 2 5 35 1	35 10 39 40 36 39 31 5	10 19 2 6 34 39 30 31	28 15 4 38 7 3 19 13 9	28 3 2 10 22 35 26 32	31 1 3 23 13 14 4 1 35	35 20 34 1 31 38	12 8 18 13 35 25 22 21 31	15 29 30 35 1 13 4 19	13 15 6 10 33 1 28	15 3 26 25 31 13 30 12 1
8 Volume of Stationary Object		1 7 28 5 2 19 12 37	35 1 39 18	10 35 18 1 28 32 16	1 35 40 5 30 7	7 32 14 24 17 26 3	31 39 14 3 17 35 4 13	31 35 5 1 31 30 2 21 31	4 30 2 21 31	31 28 6 29 2 13 32	28 24 2 7 17 40	7 26 24 17 1 15 28
9 Shape		35 6 13 40 3 31	29 35 30 3 5 24 2	10 28 5 34 24 17 26	4 14 15 3 31 25	17 2 7 15 28 32	9 4 35 30 14 3 28	35 1 17 7 19 24	1 35 30 31 22 13 5 12	35 1 28 29 15 3 31	28 3 31 10 24	32 26 25 37 15 3 29
10 Amount of Substance		3 30 1 35 38 24 18	24 4 10 34 3 12 6 17	3 25 19 1 16 38 18	35 7 18 19 24 38 13	15 19 7 32 4 37 1	31 10 9 1 4 31	35 1 77 19 24	35 40 3 12 10 19 36 39 25	1 15 17 29 24 3	35 2 24 33 21 30	25 10 17 6 19 13
11 Amount of Information		27 19 28 3 13	2 13 7 28 13 17	2 7 3 19 13 28 17	2 10 36 28 24 38 13		3 4 2 9 37 10	28 2 10 5 3	2 10 13 17 31 28 32	1 15 17 29 25 31	6 25 10 24 13	25 10 17 6 19 13
Performance												
12 Duration of Action of Moving Object		13 1 19 12 3 29 28 15	18 13 14 28 3 34 10 38	28 3 19 10 24 5 16	35 7 18 19 13	15 19 7 32 4 37 1	17 7 18 9 16 24	40 3 37 6 11 30 4 39 14	40 3 37 6 11 33 13	13 5 4 17 2 35 40	10 24 17 28 11 3	25 10 24 12 1 26 37
13 Duration of Action of Stationary Object		3 19 12 15 34 2 5 35 1	35 10 17 4 7 18 16 4	3 24 10 14 5 28 16	10 12 25 40 14 24 19	10 7 2 24 25 19 35	31 35 24 17 5 30	3 14 35 4 1 18 13	35 16 40 3 39 13 33	13 5 4 17 37	26 28 15 13 11 3	25 1 3 10 37 13 24
14 Speed		35 13 10 04 19 3 4 1 20	36 35 24 4 7 18 16 4	28 31 9 10 24 5 28 16	19 1 35 14 24 19 5 13	37 28 2 7 26 37 3	31 35 40 14 4 29 9	35 21 2 21 21 31 19	35 21 2 21 33 28 1 16	15 10 26 3 1 30 35	7 19 6 24 3 2	28 37 13 25 32 15 17
15 Force/Torque		19 35 6 13 34 15	3 35 40 12 17 5 28	10 3 23 37 14 35 34	19 15 2 5 14 35 34	37 32 7 24 1 13 10	13 9 24 12 10 5 7 14	15 2 35 5 21 3 13	35 13 24 14 17 23 40 12	15 17 3 19 29 15 4 18 24	28 5 6 25 24	1 28 25 4 3 1 29 17 38
16 Energy Used by Moving Object		2 13 28 12 10 15 3 6	3 35 40 12 9 34	10 18 28 5 35 38 20	19 15 2 24 21 24 13	37 32 24 13 1 13 40	17 3 25 17 4 31	1 24 19 21 4 15 2	35 2 11 16 6 36 28 3	35 13 17 1 30 7	10 24 28 1 13	28 3 35 1 11 3
17 Energy Used by Stationary Object		2 19 15 13 12 28 25	30 1 28 29 6 9 34 27	2 7 3 19 28 17	15 19 7 32 4 37 1	2 10 7 13 3	31 34 23 3 3 41	21 4 22 42 1 18 13	1 8 2 35 19 4 28 12	35 28 31 40 1	24 3 35 12 33 13 28	13 15 24 10 11 3
18 Power		2 28 14 15 3 6 24	31 28 18 12 4 9 34	10 15 36 19 24 16	10 2 25 40 19 24 16	37 28 27 24 19 39 35	34 28 23 13 5 30	1 3 35 19 24 10 25	1 18 2 35 2 33 25 35 39	1 35 28 15 35 12 34	24 28 15 13 12 11 7 2	26 10 35 25 24 2 37 12
19 Stress/Pressure		2 40 13 17 5 18 8 8	10 25 3 4 17 37 24 36	14 37 13 10 26	3 25 2 35 12 17 30	37 28 2 7 26 37 3	35 40 27 4 29 9	35 24 2 1 21 35	35 21 2 33 18 17 31 34	15 35 16 35 30 1 10 35	25 10 6 28 4 17	17 35 40 2 26 11 14
20 Strength		1 19 17 13 31 40	3 35 40 12 17 5 28	10 3 23 37 14 35 34	19 15 2 24 21 24 13	37 32 7 24 1 13 40	3 18 24 4 31	15 24 21 1 21 35	34 39 19 27 25 31 37 30	40 35 16 35 4 1 29 15	4 12 28 13 7 14 24	17 35 40 2 25 24 1 10
21 Stability		10 3 12 24 5 15	35 40 12 17 5 28	19 9 35 3 5 29 28	19 15 2 24 21 34 36	32 13 4 1 14 6 25	3 19 15 4 31	19 22 31 2 18 35	35 2 22 3 10 3 12 20	35 19 1 42 13 29 2 10	1 6 12 28 7 14 24	28 3 3 35 1 26 19
22 Temperature		3 24 10 16 19 13 4	29 21 3 36 31 34 13 37	19 9 35 21 28 37	35 21 24 5 3 5 19	32 13 4 19 35	35 13 39 18 24 31	3 19 15 4 18 35	35 32 19 3 19 40 13 37 17	15 35 4 40 12 35	24 2 13 30 15 17 3	28 24 26 2 23 15 45 3
23 Illumination Intensity		35 13 19 31 40	13 1 15 35 13 16	19 1 26 35 19 17 4	19 1 35 32 13 6 14	32 13 14 6 25	35 28 39 2 24 31	35 19 40 2 19 18 35	35 32 1 32 35 40 13 37 17	15 35 4 40 12 35 31	32 35 5 13	28 4 35 24 19 12 5 26

40 Inventive Principles

#	Principle	#	Principle
1	Segmentation	21	Skipping
2	Taking Out/Separation	22	'Blessing In Disguise'
3	Local Quality	23	Feedback
4	Asymmetry	24	Intermediary
5	Merging	25	Self-Service
6	Universality	26	Copying
7	'Nested Doll'	27	Cheap Short-Living Objects
8	Anti-Weight	28	Mechanics Substitution
9	Preliminary Anti-Action	29	Pneumatics And Hydraulics
10	Preliminary Action	30	Flexible Shells And Thin Films
11	Beforehand Cushioning	31	Porous Materials
12	Equipotentiality	32	Colour Changes
13	'The Other Way Around'	33	Homogeneity
14	Curvature	34	Discarding And Recovering
15	Dynamization	35	Parameter Changes
16	Partial Or Excessive Actions	36	Phase Transitions
17	Another Dimension	37	Thermal Expansion
18	Mechanical Vibration	38	Strong Oxidants
19	Periodic Action	39	Inert Atmosphere
20	Continuity Of Useful Action	40	Composite Materials

Enhanced Performance (Contradiction Matrix, parameters 35–40)

Column parameters: 35 Reliability/Robustness · 36 Repairability · 37 Security · 38 Safety/Vulnerability · 39 Aesthetics/Appearance · 40 Other Harmful Effects Acting on System

Worsening ↓ / Improving →	35	36	37	38	39	40
35	3 17 1 35 / 27 14 11 4 / 31	2 28 27 / 3 30 33 11 / 14	2 5 6 23 24 / 28	26 10 17 13 / 35 2 5 28	3 1 31 14 / 40 30	5 24 27 21 / 2 25 18 31 / 35
36	10 2 14 40 / 28 29 25	2 17 13 27 / 11	28 2 17 30 / 4 13 14	40 9 31 3 7 / 14	7 31 3 4 32 / 1	17 37 37 3 / 19 9 4 34
37	35 10 14 17 / 40 3 29 2 1	1 28 10 17 / 27 2 13 25	17 7 3 28 / 13 24	17 14 30 35 / 29 28 19 30	17 14 3 32 / 4 5	1 17 15 25 / 24 10 35 28 / 19
38	35 17 31 29 / 28 12 40	17 13 1 2 / 35 22	28 9 26 / 17	35 24 3 12 / 31 39	3 17 32 7 / 35 22	1 18 9 16 / 35 40 39
39	17 3 28 35 / 29 5 2 9	3 10 17 1 / 37 35 16	1 28 10 35 / 2 19	3 35 14 29 / 19 1 5	17 3 14 15 / 13 4 5 7 1	28 1 3 2 33 / 13 22 27 35
40	35 40 4 5 / 25 28 7 32	1 16 32 17 / 13	2 28 24 10 / 1	31 40 35 4 / 24 17 1	14 3 32 26 / 4 1	35 39 24 1 / 2 33 37 12 / 3

40 Inventive Principles

#	Principle	#	Principle
1	Segmentation	21	Skipping
2	Taking Out/Separation	22	'Blessing In Disguise'
3	Local Quality	23	Feedback
4	Asymmetry	24	Intermediary
5	Merging	25	Self-Service
6	Universality	26	Copying
7	'Nested Doll'	27	Cheap Short-Living Objects
8	Anti-Weight	28	Mechanics Substitution
9	Preliminary Anti-Action	29	Pneumatics And Hydraulics
10	Preliminary Action	30	Flexible Shells And Thin Films
11	Beforehand Cushioning	31	Porous Materials
12	Equipotentiality	32	Colour Changes
13	'The Other Way Around'	33	Homogeneity
14	Curvature	34	Discarding And Recovering
15	Dynamization	35	Parameter Changes
16	Partial Or Excessive Actions	36	Phase Transitions
17	Another Dimension	37	Thermal Expansion
18	Mechanical Vibration	38	Strong Oxidants
19	Periodic Action	39	Inert Atmosphere
20	Continuity Of Useful Action	40	Composite Materials

Contradiction Table (Worsening Features 41–48)

Worsening feature columns — Manufacture/Cost: 41 Manufacturability, 42 Manufacture Precision, 43 Automation, 44 Productivity, 45 System Complexity, 46 Control Complexity. Measurement: 47 Ability to Detect/Measure, 48 Measurement Precision.

Improving features (rows): Physical (1–8), Performance (9–23).

Improving Feature	41	42	43	44	45	46	47	48
1 Weight of Moving Object	1 5 35 16 / 14 27 24 3	29 5 26 35 / 25 1 18 24 3	35 26 2 10 / 25 24 1 19	35 24 10 28 / 12 1 37 3 16	26 35 45 30 / 36 2 10 25 28	10 12 2 19 / 15 5	26 28 32 5 / 20 36 32 37	28 26 35 10 / 2 37
2 Weight of Stationary Object	35 9 9 1 10 / 14 27 24 3	10 35 17 29 / 7 16 25	2 26 30 12 / 23 19	28 35 19 4 / 29 1 3 19 25	13 4 17 14 / 26 10 1 9	17 25 13 19 / 10 3 2	17 25 37 32 / 28 15 18	28 28 18 37 / 4 3
3 Length/Angle of Moving Object	1 24 4 10 29 / 37 25	30 32 35 2 / 29 28 24	17 3 24 26 5 / 16 25 2	17 4 14 28 1 / 29 25 3 5	1 19 24 26 / 5 28 29	17 25 13 / 10 3 2	26 35 1 24 / 32 40	10 32 1 24 / 28 39
4 Length/Angle of Stationary Object	17 3 15 13 4 / 31 10	30 32 10 3 / 29 28 24	36 37 16 17	17 4 7 30 37 / 35 16	19 24 26 35 1 / 13	35 1 28 / 29 3	29 3 26 3 / 25 2	28 10 26 32 / 3 30 24
5 Area of Moving Object	3 1 13 24 26 / 35 16 20	35 2 13 25 / 14 32 1 5	14 29 1 5 30 / 23 3 2	10 2 26 17 / 15 19 34 38	14 17 5 7 35 / 1 29 16 10	35 3 5 5 / 28	2 26 19 3 / 32 36 18 31	3 32 26 25 1 / 10 37
6 Area of Stationary Object	16 17 40 13 / 10 5 36 32	35 2 29 38 / 18 32 24 25	25 10 3 28 / 13 16	10 17 7 15 / 13 15 17 20	1 26 28 18 / 36 13 17	25 10 3 19 / 23 5 24	32 35 18 2 / 30 24 28	3 28 32 26 / 37 18
7 Volume of Moving Object	1 10 3 24 40 / 29 34 13	25 28 18 9 2 / 30 23 15	24 3 35 34 / 13	2 10 35 34 6 / 13 15 17 20	2 35 26 15 / 29 5	1 25 10 2 / 23 5 24	26 24 4 4 / 35 28	26 28 24 13 / 36 5
8 Volume of Stationary Object	35 30 13 10 / 29 34 13	35 10 / 24 26 1 9 16	10 1 13 24 / 25 17	35 37 10 2 / 7	29 3 35 / 35	28 37 3 / 10 32	17 28 28 32 / 31 2	10 32 35 25 / 2 24 26
9 Shape	17 32 1 28 / 25 35	30 32 40 22 / 2 25 38	1 28 10 / 25 17	1 13 3 36 34 / 12 35 1	29 2 5 22 28 / 16	28 37 24 7 / 31 2 25	28 37 13 1 / 15 24 7	28 1 37 32 / 26 3
10 Amount of Substance	10 2 35 1 27 / 24 17 4 36	30 3 35 25 / 28 13 37	35 1 10 8 9 / 13 37	1 13 3 35 34 / 13 24	3 1 10 7 27 / 19	28 7 / 3 1 2 25	28 18 32 24 / 37 4 19	4 24 37 13 / 19 2 28
11 Amount of Information	25 13 10 1 / 2 33	13 3 10 37 27	25 1 6 13 26 / 25	2 25 19 3 10 / 13 37	2 16 5 13 / 25 40	3 25 40 10 9 7 / 2 4 5	3 4 37 25 / 46 2	37 3 17 28 / 24 4 13
12 Duration of Action of Moving Object	35 4 10 3 2 / 5 40	3 16 40 10 / 37 12 25	10 36 17 6 / 24 31 1 13	10 36 16 3 / 14 17 15 40	5 15 10 4 2 / 28 29 25	35 29 26 / 37 4	10 37 4 32 / 19 35 39 28	3 17 28 24 4 13 / 26
13 Duration of Action of Stationary Object	35 10 40 5 / 13 2	3 18 40 10 / 37 2 25	10 13 17 1 13 / 24 31	10 35 16 3 / 3 20 35	5 10 2 25 4 / 17 14	5 28 26 20 / 37 4	19 35 39 28 / 28 29 36	3 17 26 / 24 10
14 Speed	35 30 13 2 / 28 10 40 5	2 3 10 12 13 / 16 13	10 28 2 5 / 27 17 24	1 10 6 35 34 / 7	34 28 35 23 / 2 34 13 28	1 28 25 4 / 6	3 26 34 16 / 37 39 28	28 37 11 13 / 25
15 Force/Torque	4 8 1 10 2 / 17	37 10 3 2 13 / 17	10 28 8 8 34 35 / 24	28 1 10 37 24 / 35 38	28 4 3 13 / 16 34	2 25 18 10 / 30 29	28 19 35 28 / 28 24 27	28 37 4 18 / 24
16 Energy Used by Moving Object	10 28 28 21 / 34 28 24	2 3 10 35 / 15 12	2 10 13 8 / 27 10	35 35 20 34 25 / 34 23	35 30 13 28 / 2 34 24	2 35 16 35 / 28 26 30	29 35 3 5 / 16 32 37 25	37 1 35 21 / 10 32
17 Energy Used by Stationary Object	36 23 / 17	2 35 1 6 / 32	10 28 2 / 17 24	15 17 30 18 / 20	35 37 20 10 / 2 34 3 28	35 32 30 34 / 34 39 10 18	37 1 25 27 / 28	25 13 37 4 / 17 26 34
18 Power	10 28 23 21 / 34 26 8 2	2 3 10 6 23 29 / 28	2 10 13 6 / 28 11 5 29	35 34 16 / 37 35 24	2 35 34 27 42 / 3 6 3 28	32 28 2 6 / 28 26 30	29 35 10 14 / 28 10 26 35	35 17 2 40 / 3 29
19 Stress/Pressure	2 36 17 16 / 31 8	2 3 10 / 32	12 2 35 / 17 24	10 40 25 24 / 14 35 37	35 19 / 19	2 35 32 40 2 / 28 37	35 24 7 3 / 26 3	3 37 28 25 / 32 29
20 Strength	35 15 3 40 / 14 4 37 24	3 6 23 2 29 / 28 25	15 2 35 10 / 17 13	35 5 19 17 10 / 14 35 37	2 5 15 18 / 14 29 5 3 25	2 5 10 24 / 28 7	2 40 32 28 / 28 24 4	25 13 37 4 / 17 28
21 Stability	25 35 24 3 / 15 5 19	3 25 2 12 5 / 15 30	24 1 10 16 8 / 14	35 18 10 13 / 12 13	13 31 2 10 / 40 35 36 7	25 10 24 / 28 5 37	2 4 17 26 / 28 5 37	25 15 37 4 / 17 28 24
22 Temperature	35 3 28 30 / 16 21 37 5	3 35 32 39 / 24	2 28 3 19 / 12	3 15 26 28 / 40 12 7 24	13 32 2 10 / 40 35 26 7 / 17 5	32 28 35 36 / 32 40 4	32 28 19 3 / 26 7 31 37	32 19 28 26 / 37 32 28
23 Illumination Intensity	35 28 20 33 / 16 26	3 35 32 39 / 37 24	10 2 28 / 24	2 3 15 5 28 / 37	6 15 3 28 / 17 4	15 3 37 4 5 / 17 4	32 15 1 9 37	32 10 37 4 3 / 15

Physical

Worsening Feature (→, columns 1–14) / **Improving Feature** (rows 24–40)

Column (Worsening Feature) key:

1. Weight of Moving Object
2. Weight of Stationary Object
3. Length/Angle of Moving Object
4. Length/Angle of Stationary Object
5. Area of Moving Object
6. Area of Stationary Object
7. Volume of Moving Object
8. Volume of Stationary Object
9. Shape
10. Amount of Substance
11. Amount of Information
12. Duration of Action of Moving Object
13. Duration of Action of Stationary Object
14. Speed

Efficiency

Improving Feature	1	2	3	4	5	6	7	8	9	10	11	12	13	14
24 Function Efficiency	3 30 8 40 1 7 35	31 3 35 40 7 30 1	17 3 4 15 14 30 35	17 4 13 14 7 5	15 30 17 34 35 14	14 17 4 7 28 26	15 19 14 4 3 13	28 4 37 35 13 12	4 14 3 30 19 31 17	31 3 30 19 18 4 1	2 37 3 17 4 13 19	10 4 3 18 28 9 36 1	35 24 28 10 3 30 18	3 4 15 30 29 28 13
25 Loss of Substance	28 35 40 8 6 31 7 4	35 4 31 6 13 2 40	14 12 10 24 17 4	17 28 24 10 4 3	10 17 30 12 5 31 2 35	18 10 5 30 4 13 17 34 24	30 1 29 36 12 13 3 4	3 18 30 39 30 9 12 13	35 5 29 3 28 30 7 36	24 3 10 6 35 12	28 24 10 1 37 25 4	3 15 19 28 18 35 17	24 15 18 38 17 35 28	28 19 13 25 10 38 3 24
26 Loss of Time	10 20 8 14 35 37 5	10 20 26 35 5	17 13 15 29 14 7 2	5 14 17 24 13 30 2	17 15 16 5 26 4 13	35 17 4 10 5 12 16	20 10 5 18 34 2	35 5 10 12 16 3 7	7 28 17 4 10 34 9 37	35 10 5 18 16 19 38 4	2 3 10 25 5 7	3 17 28 18 15 10	5 28 24 7 16	28 26 3 10 4 5
27 Loss of Energy	8 15 28 19 6 31 14	19 7 6 31 18 14	3 4 13 17 7 28 6	4 13 7 25 17 38 6	15 17 30 28 26 4 13	17 14 7 38 18 30 12 25	7 5 28 4 18 19 3	7 5 3 4 19 28 12	4 3 24 30 19 13 17 5	7 25 15 18 3 12 23	2 10 24 25 7	21 35 34 14 3 19 18	17 31 35 34 14 3 19	3 35 14 28 10 13
28 Loss of Information	13 35 7 10 17 24	13 35 31 5 24 10	28 25 10 37 1 26 16	28 25 17 26 37 32	28 26 17 25 30 16 1 37	28 25 26 16 37 30 17 32	28 25 17 32 7 31 1 2	28 25 32 1 35 31 7 2	25 14 24 3 4 32	31 9 14 35 4 1 10 15 39	3 10 23 31 34 2	10 28 37 3 4	10 4 1 13 24 3 35	12 13 24 26 37 32
29 Noise	31 9 3 22 13 14 4	31 9 14 39 4 35	17 3 14 9 1 35 13	3 17 14 35 9 1	3 14 17 35 1 13 9	17 3 9 14 1 35	3 14 35 1 13 4 9 17	3 9 14 4 35 1	4 35 28 2 14 22 31 3	3 19 14 35 4 1 10 15 39	3 10 23 37 13 4	10 1 35 19 21	37	3 1 14 31 39 24 4
30 Harmful Emissions	28 35 20 21 19 5 18	28 31 13 5 2 36 35	14 18 4 19 36 15 13	17 18 21 14 1 36 13	4 14 19 15 18 13 3 36	4 14 13 24 17 18 36 3	13 15 3 19 18 4 1	36 24 35 1	35 36 3 31 38 15 13	35 10 19 18 24 13 7	10 23 37 1 13 4	15 3 12 21 33 35	1 3 10 15 18 36	35 28 13 21 3 18 36
31 Other Harmful Effects Generated by System	35 30 31 15 19 40 39	40 35 31 9 1 4 30	15 17 40 24 16 35 12 3	17 4 35 40 14 24	4 17 12 3 19 7 24	3 4 1 35 5 40 17 18	17 30 14 2 40 28 1	4 14 30 18 35 5 24	1 35 4 3 7 30 17 24	2 14 35 39 5 9	10 7 32 4 3 17	15 3 5 12 21 33 35	21 16 17 14 13 9 35	10 14 35 24 15 28 12 29

Enhanced Performance

Improving Feature	1	2	3	4	5	6	7	8	9	10	11	12	13	14
32 Adaptability/Versatility	35 1 8 31 30 6 14 13 29	35 1 7 12 31 13 16	17 7 30 15 26 19 29 4 24	17 4 26 28 31	13 29 17 31 28 14 16 3	31 6 26 24 3	30 35 31 6 24 4	24 15 31 16 1 3 35 7 30	15 2 24 30 7 30 17 13	30 31 35 3 15 24 25	2 19 7 16 37 4 1	28 29 35 13 1 24 19 12	15 13 16 2 17 3 35	10 4 35 24 15 28 12 29
33 Compatibility/Connectability	1 8 15 28 26 13	1 28 15 26 13	24 4 28 17 3 15 2	24 28 17 4 2	28 17 13 1 35 2	28 13 17 1 2	2 24 28 3 26 1	2 26 24 28 1	13 28 7 24 17	28 31 2 13 26 35 5 17	10 24 3 2 37 6	2 10 7 5 40 19	2 10 7 5 40 3	1 2 10 6 5 4
34 Trainability/Operability/Controllability	28 35 25 15 13 2 26 8	28 35 25 26 13 1	17 13 1 12 3 4 15	17 1 13 4 28 3	28 26 13 17 1 3 14 16	28 26 17 1 3 16 15	1 15 35 7 16 4 13 28	28 31 2 4 19 39 17	28 7 2 3 35 32 4 13	35 1 13 17 14 4 12 2	1 7 2 10 4 17 32	19 25 3 35 5 12 7	1 16 25 3 5 35 12	13 25 5 2 24 35 28
35 Reliability/Robustness	40 5 3 12 8 28 35 31	3 35 14 28 10 8 5 40	14 17 15 4 35 9 40 3	14 24 3 35 7 4 17	17 14 15 10 30 32	17 14 1 4 3 32 24	14 7 15 24 35 3 10 1	5 35 3 17 14 24 7 31 2	35 1 40 4 30 2 17 24	3 28 40 31 25 4 2	24 4 32 10 25 5 2	3 5 2 40 25 19 13	3 35 1 11 4 16 25	35 28 24 4 10
36 Repairability	8 35 13 17 28 30	3 31 35 28 26 22	17 1 28 29 13 10 3 18	17 3 1 18 13 28 31 21	13 15 17 1 18 32	17 7 28 39 4 13	13 28 35 30 2 13 1	13 28 15 39 1 17	3 2 17 13 24 40 26	3 35 1 35 10 2 24	24 9 2 37 26 1	28 34 35 3 11 1 25	3 35 1 11 4 16 25	34 9 5 15 3
37 Security	28 30 3 22 13 35 26	3 31 35 28 26 22	17 13 28 4 30	17 28 14 29 26	17 28 1 4 13 35 2	17 14 13 4 32 24	13 28 30 1 17	13 28 15 39 1 17	3 2 17 13 40 26	13 28 35 37 2	3 28 32 24 13 2	13 14 10 37 35 17 18	35 10 37 17 18 12	28 37 4 17 3 18
38 Safety/Vulnerability	8 31 30 13 12 40 26	31 30 13 17 40 1 26	3 17 2 31 31 30 37	3 17 2 31 4 30 13	17 15 13 4 30 14 3	17 14 13 4 32 24	13 31 15 17 32 3 2	13 35 31 17 2	15 2 32 31 13 5 1	35 31 13 9 5 30	10 25 3 2 5 24 22	10 13 5 15 19 2 35	19 10 13 5 37 2 35	14 31 13 3 17 19 7
39 Aesthetics/Appearance	30 40 3 35 29 8	35 40 3 8 17	17 14 3 32 15 7	17 14 15 3 32	14 17 4 15 7 30 32	14 17 1 4 3 32 24	14 15 7 28 32 3 2	14 13 4 32 7 24 2	15 2 32 13 5 1	30 31 40 3 35 29 1 28 2	7 24 17 10 28 32 2 3	3 2 6 32 11 28	2 28 6 32 15 1 5	15 3 14 19 26
40 Other Harmful Effects Acting on System	35 31 8 21 3 30 40	35 2 31 13 40 24 3 17	17 14 3 32 15 7	17 28 14 29 26	17 28 1 4 15 30 32	17 7 14 13 32 24 39	3 24 15 37 35 7 7 19	5 17 39 19 4 35	30 24 32 1 35 17 13	31 30 35 28 35 29 1 28 2	2 26 3 3 40 25 17	15 28 35 24 21 4 3	35 5 40 3 1 24 33	24 35 28 21 13 3 19 1

40 Inventive Principles

No.	Principle	No.	Principle
1	Segmentation	21	Skipping
2	Taking Out/Separation	22	'Blessing In Disguise'
3	Local Quality	23	Feedback
4	Asymmetry	24	Intermediary
5	Merging	25	Self-Service
6	Universality	26	Copying
7	'Nested Doll'	27	Cheap Short-Living Objects
8	Anti-Weight	28	Mechanics Substitution
9	Preliminary Anti-Action	29	Pneumatics And Hydraulics
10	Preliminary Action	30	Flexible Shells And Thin Films
11	Beforehand Cushioning	31	Porous Materials
12	Equipotentiality	32	Colour Changes
13	'The Other Way Around'	33	Homogeneity
14	Curvature	34	Discarding And Recovering
15	Dynamization	35	Parameter Changes
16	Partial Or Excessive Actions	36	Phase Transitions
17	Another Dimension	37	Thermal Expansion
18	Mechanical Vibration	38	Strong Oxidants
19	Periodic Action	39	Inert Atmosphere
20	Continuity Of Useful Action	40	Composite Materials

Performance

15 Force/Torque	16 Energy Used by Moving Object	17 Energy Used by Stationary Object	18 Power	19 Stress/Pressure	20 Strength	21 Stability	22 Temperature	23 Illumination Intensity
35 40 17 13 3 9 19 7	2 4 15 3 19 35 17	3 35 19 15 17 14 4	35 3 15 19 17 4 38	3 17 35 31 19 12 40 15 9	35 40 3 4 19 12 15 9	35 2 19 30 9 17 38 15 3	19 35 3 31 28 21 37 24	19 3 35 28 32 5 31 17
14 15 9 18 40 35 17 4	18 35 5 3 19 12 28	12 18 28 35 30 24 31 19	28 18 38 25 13 3	3 37 10 1 17 36 9 12	35 28 3 40 17 31 4 34	1 30 19 24 29 36 18	36 37 21 39 31 24 2	2 13 35 28 6 1 24
5 17 10 37 36 3	13 35 38 3 4 19	35 10 1 19 38 3 4	35 6 10 1 20 12 24	35 17 4 20 36 37 9	35 24 3 9 28 18 4 29	24 5 3 35 17 31 28	21 18 24 35 31 3	26 1 32 17 13 19 2
19 17 2 36 4 38	35 19 3 4 37 2	35 19 4 2 12 34 3	19 4 37 34 38 21	2 25 4 13 12 19	28 35 17 26 31 40 9 37	10 6 14 12 2 19 28	35 7 31 34 19 21 1	19 24 5 13 35 32 15
13 17 24 37 1 36	1 24 25 20 10 19	10 1 24 36 25	10 19 24 25 7 13 3 37	24 22 25 35 7 14 31	35 24 3 31 14 40	35 30 10 26 24 5 40	28 22 1 25 26 24	19 32 24 28 3 5
3 14 17 4 1 31	19 28 4 35 14 24 23 9 3	19 23 28 4 24 14 9 3 35	28 23 25 24 3 13 14 35 39	3 14 9 2 23 24 39 25	3 35 26 40 4 28 30 10	23 25 2 10 13 5 29 39 12	35 12 2 13	35 2 9 40 17 13
10 3 15 35 28	35 28 10 3 20 4	35 3 20 10 28 4	35 28 10 3 4 18 20	10 12 9 35 15 28 17 13	10 35 15 28 17 4 5	15 4 19 3 1	19 18 36 20 3 1 35	32 24 35 2 5 19
28 35 15 29 40 1 3 4	35 6 12 26 4 20 4	35 24 19 3 4 22	3 35 18 4 14 13	37 17 4 3 9 14	35 40 17 5 30 2 15	40 4 35 14 24 3 39 14	35 24 1 7 5 3 10 13 22	19 35 32 24 39 28 1
35 15 17 14 6 7 13	29 13 19 35 15 16 12 1	16 1 19 3 12 17	19 1 24 35 29 28 12	15 23 13 28 24 12 4 16	35 40 3 17 13 24 9 19	35 40 4 14 30 21 24 3	35 5 19 36 2 22 7	1 35 32 19 17 24 28 26
6 24 29 12 3 15	29 28 12 3 25 2 13	2 24 25 19 5 13	6 29 25 28 12 16 2	24 40 3 12 9 5 13 29	2 35 29 30 17 9 33	35 24 33 27 3 10	21 39 35 9 22 7	25 28 2 6 13
28 13 24 31 34 17	24 1 13 28 12 19 3	24 28 12 3 13 15	1 10 28 35 2 21 36 37	1 12 23 4 29 2 9 35	19 3 12 16 1 4 7	25 1 30 24 40 19 3	31 13 24 26 35 19	19 13 24 1 17 32
8 28 3 4 1 17 14 9	35 9 13 14 21 37	35 19 3 1 13 12	35 1 4 10 40 16 3	35 10 19 24 40 3 5 12	35 40 3 4 12 1 24 28	40 35 3 1 39 24 2	3 35 15 10 30 37 36 1	35 37 24 13 32 3 11 5
1 10 7 15 13 31	28 1 15 13 3 16	28 13 3 1 16	15 10 2 13 1 4	1 3 13 25 31 4 40 9	1 4 17 9 24 29 3 37	2 35 7 19 24 25	24 13 4 37 10 29 25 36	15 13 32 1 3
2 3 17 14 13 9 29	32 12 13 25 4 9	32 12 4 9 13	12 13 9 35 37 32 25	37 17 4 3 9 14	28 2 13 17 40 14	13 12 5 24 3 17	1 31 37 19 4 3 7	28 37 26 32 3 1
17 13 19 3 14 7 31 24	35 12 19 1 39 24 5	39 35 30 4 1 19	1 19 34 11 13 24 23	35 40 3 9 31 4 5	35 31 40 3 9 7 4 5	35 5 40 31 3 39	31 35 36 3 19 2 13 4	28 19 24 2 13 35
3 28 7 31 4 15 19 14	15 3 19 28 7 4 14 1	3 28 19 14 1 8 38	15 4 14 32 1 19 24	40 9 17 7 5 2 26 35	9 17 40 2 35 7 5 3	3 40 10 35 4 31 29 17	35 31 3 15 2 36 40	32 3 35 19 1 5 15 4
3 13 35 18 17 24 33 39	6 24 1 26 15 14 17 3	1 35 24 6 26 17 3 14	19 2 31 10 24 6 1	1 35 40 2 14 12 4 30	35 1 40 17 12 3 5 14	35 24 4 40 30 3 39 28	35 31 33 17 12 40 2 5	35 1 13 32 19 5 40 28 25

Worsening Feature →

Improving Feature ↓

Efficiency (worsening features 24–31)

Improving Feature	24 Function Efficiency	25 Loss of Substance	26 Loss of Time	27 Loss of Energy	28 Loss of Information	29 Noise	30 Harmful Emissions	31 Other Harmful Effects Generated by System	32 Adaptability/Versatility	33 Compatibility/Connectability	34 Transability/Operability/Controllability
Efficiency											
24 Function Efficiency	28 10 3 13 4 / 1 18 15	4 12 17 34 2 / 24 30 3	3 14 28 15 / 24 12 1	3 4 31 19 15 / 28 2 38	3 4 19 15 32 / 17 13 31	14 9 3 4 1 2 / 15	35 24 18 28 / 22 13 4	19 15 4 33 3 / 37 30 39	15 19 3 28 4 / 30 16	1 4 14 7 2 / 24	25 10 1 13 3 / 19
25 Loss of Substance	5 19 3 20 16 / 34	░ shaded	35 15 2 18 / 28 10 24 36	19 38 35 27 / 34 36 5 4	32 40 1 10 3	35 17 7 2 28 / 4	13 2 24 35 / 28 21	3 1 15 14 29 / 13 12 9	2 15 28 27 / 12 15 3	34 2 33 24 / 13 5 15	3 15 32 2 1 / 14 24 34
26 Loss of Time	1 13 19 4 12 / 37 16	24 10 35 18 / 4 12 17 13	░ shaded	5 18 35 19 1 / 4 13	24 28 32 2 / 26 5	9 31 14 18 / 19 4 24	1 35 21 18 5	35 24 14 39 / 22 18 6	28 35 40 13 / 17 4	17 10 28 2 / 24 1	10 25 4 26 / 28 1
27 Loss of Energy	10 5 37 12 / 32 25	28 35 37 2 / 27 3 4	10 18 7 4 32 / 15	░ shaded	19 10 4 37 1 / 7	28 35 14 25 / 2 9	2 1 21 19 3 / 35	40 35 21 24 / 2 28 4 5	15 5 14 13 / 31 35 3	21 35 12 13 / 17 4 24	13 1 35 25 / 10 32 4
28 Loss of Information	19 10 4 37 1 / 7	7 17 2 3 13 / 25	25 22 2 3 13 / 26 28	15 9 31 35 / 39 24 13	░ shaded	2 37 3 10 25	7 3 16 26 37 / 13 24	7 10 31 27 / 21 40 6	24 5 25 9 40 / 35 19	3 9 13 26 1 / 37 32	7 1 3 10 5 / 13
29 Noise	17 15 31 3 / 25 4 35 28	35 31 2 9 13 / 24 34	2 10 35 36 / 21 3	3 15 9 31 35 / 39 24 13	10 23 2 3 26	░ shaded	5 2 35 19 9 / 25 15 7	21 23 33 1 / 34 40 12 11	28 10 1 15 3 / 25	2 35 9 17 28 / 3	28 26 25 1 / 24
30 Harmful Emissions	10 35 28 15 / 12 3	10 12 18 14 / 13 1 3 4 35	10 18 3 25 / 36	10 18 32 7 / 13	9 16 13 2 4 / 23	18 35 35 14 / 31 16	░ shaded	24 25 4 30 / 35 17 12 11	18 35 10 13 / 15 1 4	3 2 35 19 5 / 3 4	25 10 13 26 / 1
31 Other Harmful Effects Generated by System	35 28 25 2 / 15 19 3 4 13	35 31 2 9 13 / 24 34	35 28 19 25 / 37 15 13	3 15 9 31 35 / 39 24 13	7 3 16 26 37 / 13 24	14 31 13 15 / 18 9 1 35	1 19 21 10 / 24 5 35	░ shaded	1 15 29 14 3 / 17 4 24	5 17 1 29 23 / 24 3	25 10 13 26 / 1
Enhanced Performance											
32 Adaptability/Versatility	39 3 13 25 4 / 37 35	15 10 3 15 / 19 40 24	15 28 19 35 / 1 31	15 35 12 / 36 25 4	15 9 13 26 37	2 10 24 37 / 25 31	24 25 4 30 / 35 17 12 11	28 2 18 29 / 9 25 31	░ shaded	1 5 3 28 2 / 25 13	15 24 3 4 28 / 14 13 26 10
33 Compatibility/Connectability	12 3 19 37 2 / 15	13 3 17 22 4 / 9 12	2 10 13 25 5 / 17	2 13 9 12 5	5 2 7 24 13 / 4	2 3 31 35 / 28 18 17	3 11 32 33 / 24 16 30	3 11 32 33 / 24 16 30	28 10 24 6 / 15 7	░ shaded	10 25 2 22 / 28
34 Transability/Operability/Controllability	25 10 2 37 / 19 3 32	24 2 3 4 34 / 28 12	23 10 4 32 / 28 5 25	19 25 13 3 2 / 28 16	10 37 1 24 / 32 4 26	2 3 31 9 35 / 28 18 17	25 19 2 1 20 / 3 37	35 31 24 4 / 32 25 27 1	10 25 1 26 5 / 24 29 28	25 1 24 6 28 / 4	░ shaded
35 Reliability/Robustness	15 3 19 35 / 13 28 23	35 15 10 12 / 3 4 39 2	10 2 25 5 4 / 30 3	10 19 35 23 / 28 3 40	10 25 28 32 / 5 24	35 23 25 4 9 / 28 24	3 24 13 35 / 19 31 21 1	2 26 35 40 / 33 7 4 19 3	35 28 13 12 / 24 29 19 3 4	1 25 10 3 5 / 35 13	28 1 40 29 3 / 19 13
36 Reparability	2 7 27 19 4 / 2	35 34 4 12 / 24 36	26 28 25 2 9 / 17	15 14 24 2 / 1 31	3 9 13 26 1 / 37 32	14 17 31 3 5 / 35	2 35 34 27 / 6 40	35 15 39 12 / 28 40 33	1 28 24 32 / 17 13	2 10 13 4 17 / 24	1 15 26 25 / 10
37 Security	2 1 17 3 10 / 25	35 1 5 7 24	2 10 13 25 5 / 17	26 32 13 25 / 4 9	2 37 4 16 13	37 4 2 13 35 / 3	3 5 27 25 21 / 15 19 11	3 5 27 25 21 / 15 19 11	30 13 15 28 / 17 29 5	15 17 24 4 6 / 37 1	25 10 13 3 / 22
38 Safety/Vulnerability	12 13 1 7 31 / 35 19 2	34 12 13 3 / 31 24	10 5 13 12 / 19	19 13 12 35 / 1 24 3	3 24 28 5 7 / 13	28 19 24 39 / 13 10	35 31 1 33 / 16 21 11	35 31 1 33 / 16 21 11	28 7 15 29 / 13 1 3 2	24 28 2 13 6 / 37 1	2 4 26 19 13 / 16
39 Aesthetics/Appearance	2 13 28 5 27 / 4 15 29	28 17 3 4 1 / 34 12	7 10 6 2 9 / 12 24 15	28 3 15 31 / 34 35 1	3 7 32 10 4 / 24 19	3 14 35 31 / 13 9 4	4 28 15 29 / 35 1 10 7	35 2 13 29 / 31	35 13 15 30 / 4 11 16 40	28 5 17 3 32 / 7	28 6 7 17 3 / 24
40 Other Harmful Effects Acting on System	2 3 35 28 10 / 19 5 4	40 3 4 24 12 / 34 33 35	18 3 24 23 / 35 40 10 12 / 32	35 24 21 2 / 30 17 3 4	32 1 40 25 2 / 3 31	31 1 17 14 / 35 3 7	24 35 18 1 / 32 40	35 3 13 24 / 17 4 32 40 / 18	35 15 31 30 / 4 11 16 40	17 24 2 5 7 / 3	25 28 3 15 / 10 4 6 2 31

40 Inventive Principles

1 Segmentation
2 Taking Out/Separation
3 Local Quality
4 Asymmetry
5 Merging
6 Universality
7 'Nested Doll'
8 Anti-Weight
9 Preliminary Anti-Action
10 Preliminary Action
11 Beforehand Cushioning
12 Equipotentiality
13 'The Other Way Around'
14 Curvature
15 Dynamization
16 Partial Or Excessive Actions
17 Another Dimension
18 Mechanical Vibration
19 Periodic Action
20 Continuity Of Useful Action
21 Skipping
22 'Blessing In Disguise'
23 Feedback
24 Intermediary
25 Self-Service
26 Copying
27 Cheap Short-Living Objects
28 Mechanics Substitution
29 Pneumatics And Hydraulics
30 Flexible Shells And Thin Films
31 Porous Materials
32 Colour Changes
33 Homogeneity
34 Discarding And Recovering
35 Parameter Changes
36 Phase Transitions
37 Thermal Expansion
38 Strong Oxidants
39 Inert Atmosphere
40 Composite Materials

Enhanced Performance

Improving ↓ / Worsening →	35 Robustness/Reliability	36 Reparability	37 Security	38 Safety/Vulnerability	39 Aesthetics/Appearance	40 Other Harmful Effects Acting on System
24	35 30 40 27 15 28 1 4	2 27 17 1 16 4	28 2 1 24 25 7 23	31 30 2 24 23 35	22 2 32 3 4	35 13 18 31 33 17 32 12
25	10 12 35 24 17 36 40 8	2 14 27 35 4 34 24	2 17 12 28 7 14	24 19 31 5 10 30	13 28 17 4 35 7	35 30 40 24 1 22 12 32 28
26	35 10 3 14 4 30 28	17 24 32 1 10 27	2 28 26 9 25 13 24 15	25 12 13 11 24 31	17 10 4 28 32 7 19	35 18 1 34 3 30 31 33
27	35 40 17 11 10 39 37 36	1 19 30 2 29 4 13	2 3 2 26 24 1	19 4 35 31 1 16 13	4 7 28 17 3 13	21 35 33 15 19 6 30 37 36
28	13 24 10 26 6 4 3 40 17	2 10 17 13 28	26 24 25 1 17 28	10 28 24 7 30 4	32 3 17 7 5	10 24 11 3 25 22 17 7 6
29	35 9 13 16 5 21 3	5 9 31 2 17 3	28 2 24 1	5 28 24 18 22 13	9 7 31 14 4 2 35	2 35 31 22 40 13 16 5 19
30	2 35 40 3 4 17 14	18 28 1 23 27 30	28 10 2 17 13 1	15 28 23 35 31	17 7 10 5 19	2 30 37 38 40 12 22
31	40 24 35 4 12 17 7 39 33	1 24 27 17 30 12 40	28 35 1 17 12 30	31 30 35 39 16 24 10 3 5	2 17 28 24 40 3 26 35	35 1 24 9 3 25 7 13 12
32	35 40 24 30 10 36 17 31	1 7 4 3 28 2 23	28 24 18 10 4 17 1	3 16 13 30 31 12 24 5	28 7 13 17 3	1 4 35 12 24 3 5 28
33	9 35 24 10 37 12	28 10 17 13 9 2	24 10 28 25 1 15	17 25 9 3 31 19	28 29 22 32 3 7 24	35 2 25 11 39 16 5 30
34	40 35 2 17 12 29 23	1 17 13 24 26 3 27 19	3 28 5 23 24 13 25	2 25 28 4 5 24 13 17	3 14 31 29 7 32	40 2 5 12 33 25 37 19 3
35		1 11 15 27 25 7	10 13 2 24 20 17	4 10 2 22 13 28	17 7 13 30 37	10 3 40 16 30 21 2 39
36	35 10 5 7 11 30		1 28 3 10 4	15 23 1 31 26 30	7 28 3 1 24 17	2 10 40 35 33 30 39
37	35 37 13 4 2	13 4 3 2 27		2 13 7 22 17 31	2 7 30 19 3 15 9	31 33 9 4 16 3 24
38	28 35 25 36 40 37 5	17 1 28 13 3	2 13 7 22 17 31		2 7 30 19 3 15 9	2 12 5 19 40 31 30 29
39	2 28 35 3 4 40 13	3 7 13 17 28 24 27	3 28 5 23 24 13	3 24 7 14 2 5 13 17		2 12 5 19 40 31 30 29
40	35 4 24 17 40 2 28 5 14	10 3 14 25 27 5 4	3 24 7 14 2 5 13 17	17 26 4 30	30 24 32 17 13 3 2	

| | | Manufacture/Cost | | | | | | Measurement | |
		41 Manufacturability	42 Manufacture Precision/ Consistency	43 Automation	44 Productivity	45 System Complexity	46 Control Complexity	47 Ability to Detect/ Measure	48 Measurement Precision
24	Function Efficiency	3 35 1 25 28 29 2	3 2 1 25 28 4 37 24	2 25 19 3 15	1 3 15 4 2 25	2 15 19 28 35 30 4 17	25 1 37 4 10	32 4 18 28 2 3	37 4 28 32 18 3
25	Loss of Substance	15 5 34 33 10 17 13	24 31 10 3 17 35	10 35 18 25 5 29	5 25 28 24 10 35 19 3 34	28 5 2 10 24 4 31	3 25 14 34 10 37	28 24 3 17 10 35 13 33 18	24 10 14 3 31 35 4
26	Loss of Time	10 35 4 19 34 28 3	26 25 24 18 28 4 13	10 2 24 30 5 25 12	10 3 24 5 4 13	28 21 6 10 2 12	10 37 4 5 2 3 19	37 10 18 32 28 4	25 26 32 28 24 34 4
27	Loss of Energy	10 14 35 1 29 30	30 37 9 25 3 35 29	28 1 10 2 12	3 10 35 28 29 6 2	25 7 40 30 37 4	25 37 4 2 16 12	1 3 35 15 28 37 4	3 37 24 26 32 28
28	Loss of Information	25 29 28 40 37	25 17 37 1 32 4	5 24 35 10 25	5 10 13 24 25 15	6 25 13 24 4 28	10 6 25 2 3 19	28 32 1 10 37 7	37 4 32 10 7
29	Noise	35 5 6 13 21 18	15 3 10 2 28 1	15 28 10 1 3 13	3 15 10 2 28	6 13 1 24 15 25	23 10 28 3 2 25 6 9 1	24 3 14 30 28 37 13 39	5 10 24 22 3 25 23
30	Harmful Emissions	30 13 31 5 40 10	2 18 15 25 12	10 2 36 7 14	10 2 1 19 25 21	21 28 35 6 2 10	10 23 37 1 4	24 10 28 25 4 37	28 10 37 4 3
31	Other Harmful Effects Generated by System	28 4 40 24 7 3 1	17 26 4 10 35 1 34	2 3 17 13 30 19	28 2 25 35 18 5 40 39	19 31 1 35 2 17 3	1 26 19 23 17 13 1	1 37 32 31 40 5 24 17	4 17 26 10 37 34 24
32	Adaptability/ Versatility	10 13 29 31 1 28 24 5	40 35 29 16 25 3 37 4	6 10 28 29 15 35 2	1 15 25 2 10 24 13 17	6 28 29 31 35 40 17 25	28 25 37 19 3 4 1	25 28 32 40 37 4 29	1 10 35 4 37 3 12 13
33	Compatibility/ Connectability	2 16 17 30 35 28	2 10 9 35 13 17 4 14	10 17 16 13 5 37	3 17 14 10 5 25	28 24 13 12 5 17 4	10 2 25 5 13	25 10 37 17 28 13	25 37 17 4 28 12
34	Transability/Operability/ Controllability	29 36 24 5 12 2 10 1	15 29 35 2 28 13	2 10 1 12 3 34 13 28	1 5 28 15 25 13 17	28 2 9 5 12 32 17 26	1 25 37 5 3 10	28 5 12 16 10 9	25 13 1 32 2 24 37
35	Reliability/ Robustness	28 10 35 4 40	1 13 10 4 32 39 12	35 10 1 13 17 27 28	35 1 10 28 29 32 33	5 35 13 33 15 29 3 17	1 19 25 37 10	28 40 25 32 37 18 3	3 10 37 32 7 4
36	Reparability	10 7 35 32 9 2 30	10 35 2 1 25	10 13 34 7 35 25 1	2 10 1 32 25 17 13	35 30 28 17 6 13 1	2 35 4 10 1 17 13	25 10 3 32 37 24	37 10 13 25 2 24
37	Security	10 1 12 13 17 2 35	10 3 25 16 13	2 10 17 12 25	6 28 17 10 4 19 3 32	2 6 4 17 13 26	35 37 9 26 4	37 7 4 28 13 31	37 4 26 3 12 25 2
38	Safety Vulnerability	30 31 10 3 36 18 13	10 3 25 16 13	10 3 25 13 24 15 29	10 13 1 15 8	5 31 10 27 30 40	25 7 10 1 28	28 26 31 28 17 13 6 15 30	28 37 32 37 17 3 13 26 4
39	Aesthetics/ Appearance	16 10 2 6 22 1 25 13	3 22 10 24 32 13 30	10 29 28 1 35 17	28 10 35 1 17 3	2 13 24 7 5 14 17	13 2 32 19 37 26	32 26 31 28 17 13 6 15 30	26 10 32 37 3 17 7 28
40	Other Harmful Effects Acting on System	35 24 1 40 3 39 4 13 17	2 1 10 18 26 23 12	10 3 34 4 33 1	23 35 24 13 2 40 4 9 6	4 40 2 5 17 19 29 28	1 25 37 24 9 28 26	1 17 18 32 5 40 28 39	5 37 18 32 4 28

Worsening Feature ↑ →

Improving Feature

Efficiency (24–32) — Enhanced Performance (33–40)

40 Inventive Principles

#	Principle	#	Principle
1	Segmentation	21	Skipping
2	Taking Out/Separation	22	'Blessing In Disguise'
3	Local Quality	23	Feedback
4	Asymmetry	24	Intermediary
5	Merging	25	Self-Service
6	Universality	26	Copying
7	'Nested Doll'	27	Cheap Short-Living Objects
8	Anti-Weight	28	Mechanics Substitution
9	Preliminary Anti-Action	29	Pneumatics And Hydraulics
10	Preliminary Action	30	Flexible Shells And Thin Films
11	Beforehand Cushioning	31	Porous Materials
12	Equipotentiality	32	Colour Changes
13	'The Other Way Around'	33	Homogeneity
14	Curvature	34	Discarding And Recovering
15	Dynamization	35	Parameter Changes
16	Partial Or Excessive Actions	36	Phase Transitions
17	Another Dimension	37	Thermal Expansion
18	Mechanical Vibration	38	Strong Oxidants
19	Periodic Action	39	Inert Atmosphere
20	Continuity Of Useful Action	40	Composite Materials

Physical

Improving Feature	1 Weight of Moving object	2 Weight of Stationary object	3 Length/Angle of Moving Object	4 Length/Angle of Stationary Object	5 Area of Moving Object	6 Area of Stationary Object	7 Volume of Moving Object	8 Volume of Stationary Object	9 Shape	10 Amount of Substance	11 Amount of Information	12 Duration of Action of Moving Object	13 Duration of Action of Stationary Object	14 Speed
Manufacture/Cost														
41 Manufacturability	28 1 8 15 5 / 3 29 13	1 13 14 8 26 / 24 10 27 36	14 13 1 17 / 15 10 2 29 5	13 17 14 15 / 4 29 2 27 37	13 1 26 17 / 12 4 2 16 30	1 3 16 26 13 / 40 18 30	13 1 7 30 3 / 35 40 29	1 35 15 38 / 36 2 3	29 13 1 16 / 28 30 24 27 35	35 16 1 31 / 24 30 27 29 23	6 2 10 7 15 / 13 1	10 1 37 4 21 / 12 27 13	10 1 16 35 / 37 3 13 38	35 13 1 28 2 / 8 15 4
42 Manufacture Precision/Consistency	13 8 18 28 / 16 24 26	35 28 9 17 / 31 26	3 17 28 10 / 29 37 24	17 1 10 32 / 24 37 2	29 28 37 32 / 24 33	29 18 36 37 / 32 2	28 37 1 25 / 18 20 24	35 28 25 10 / 18 20 1	30 13 10 32 / 12 21 40	30 25 32 9 / 13 12 3 31	37 3 4 32 / 17	5 40 16 3 20 / 19	5 4 17 18 3 / 30 24	10 32 28 3 / 17 4 19
43 Automation	28 13 12 35 / 18 14 31	28 12 35 10 / 13 26	17 28 13 12 / 14 4 25	17 28 13 4 / 12 6	13 17 14 12 / 4 26 5	13 26 17 4 / 12 5	26 13 35 16 / 7 24	26 13 35 31 / 16 24	13 24 10 15 / 16 1 28	26 31 35 13 / 3 23	37 4 5 24 10	19 9 16 6 13 / 10 37	25 10 16 2 / 37 13 24	10 28 25 19 / 3 37
44 Productivity	35 24 8 13 2 / 37 30 31	35 13 28 1 / 3 15	17 4 18 35 / 28 14 38	14 7 17 19 3 / 1 35 30	35 1 10 31 / 26 17 34 16	10 7 35 17 3 / 4 30	3 12 10 2 6 / 34 19 24	35 10 1 2 13 / 5 37 24 3	13 1 17 36 / 10 30 16	35 3 2 25 9 / 19 13	9 24 25 7 23 / 13 10	10 3 5 18 2 / 35 13 17	1 35 18 9 3 / 16 24 20 38	35 3 24 5 4 / 12 13 19
45 System Complexity	35 40 30 9 5 / 34 26 31 2	35 2 9 5 26 / 31 40 39	1 19 26 13 2 / 24 4	9 26 28 13 2 / 35 24 4	14 1 28 13 2 / 35 16 17 4	19 6 17 2 13 / 35 14 36	35 26 34 30 / 5 40 6 13	2 13 35 1 28 / 5	29 13 2 28 / 15 30 1	2 10 13 3 35 / 31 24	25 7 24 13 / 32 3 2 37	10 28 15 4 / 12 9 13 29	35 10 13 6 4	28 10 13 34 / 18 19 4
46 Control Complexity	10 1 6 20 31 / 35 7	10 1 13 6 35 / 7	35 17 14 10 / 4 28 12 1	28 10 25 20 / 17 13 1 5	10 13 37 5 / 28 35 31 25	10 28 2 13 / 17 37 25 4	28 37 10 25 / 19 3 5 1	28 1 10 3 5 / 2 13 25	5 25 28 10 / 13 29 37 1	10 25 7 6 35 / 37 3 19	25 10 37 7 3 / 6 13 4	3 10 37 2 5 / 6 7 12 4	10 25 2 13 / 37 28 24 7	25 28 7 10 4 / 37 5 24 19
Measurement														
47 Ability to Detect/Measure	28 26 13 5 3 / 8 35 24	3 28 26 1 13 3 / 35 6 24	26 24 28 5 / 17 3 37 16 13	28 26 10 24 / 32 2 39	26 28 13 2 / 17 32 24 18	26 28 2 17 / 39 32 24 13 9	28 24 18 1 / 32 4 13 25 26	28 26 2 24 / 13 31 32 4	13 28 3 1 17 / 26 39 24 4	3 28 18 27 / 24 13 29 4 32	19 3 32 7 10 / 13 25 4	19 26 2 13 4 / 25 39 32	26 2 35 25 3 / 6 13 24 34	3 28 1 24 37 / 25 4 16 35
48 Measurement Precision	35 26 32 1 / 12 8 25	26 1 35 8 5 / 12 10	5 26 28 1 10 / 24	26 28 10 24 / 3 32	26 24 5 3 28 / 35 10	24 5 28 3 / 35 10	5 24 28 13 / 26 10 3 1	28 24 13 3 / 35 18	28 3 10 13 / 24 1 37	2 13 1 37 6 / 24	25 2 7 32 4 / 3 37 10	10 28 6 5 34 / 26 24 27	10 26 28 24 / 5 34 27	28 13 24 5 / 32 35 37

Worsening Feature →

Improving Feature ↑

	40 Inventive Principles
1	Segmentation
2	Taking Out/Separation
3	Local Quality
4	Asymmetry
5	Merging
6	Universality
7	'Nested Doll'
8	Anti-Weight
9	Preliminary Anti-Action
10	Preliminary Action
11	Beforehand Cushioning
12	Equipotentiality
13	'The Other Way Around'
14	Curvature
15	Dynamization
16	Partial Or Excessive Actions
17	Another Dimension
18	Mechanical Vibration
19	Periodic Action
20	Continuity Of Useful Action
21	Skipping
22	'Blessing In Disguise'
23	Feedback
24	Intermediary
25	Self-Service
26	Copying
27	Cheap Short-Living Objects
28	Mechanics Substitution
29	Pneumatics And Hydraulics
30	Flexible Shells And Thin Films
31	Porous Materials
32	Colour Changes
33	Homogeneity
34	Discarding And Recovering
35	Parameter Changes
36	Phase Transitions
37	Thermal Expansion
38	Strong Oxidants
39	Inert Atmosphere
40	Composite Materials

Performance

Force/Torque 15	Energy Used by Moving Object 16	Energy Used by Stationary Object 17	Power 18	Stress/Pressure 19	Strength 20	Stability 21	Temperature 22	Illumination Intensity 23
35 12 28 29 / 1 10 3 13 2	28 1 10 26 / 35 39 19	28 1 10 26 4 / 19 35	28 12 24 19 / 1 21 10	35 10 1 19 / 21 12 13 22	3 35 1 24 33 / 10 30	1 11 3 13 39 / 33 24 9	10 24 35 18 / 2 36 26	32 24 1 35 / 28 2 27 25
12 19 28 29 / 3 10 13	2 26 32 3 19 / 21 13	2 26 21 13 / 19 3 17	2 32 21 16 3 / 19	35 3 12 13 / 17 29 32	3 17 7 35 29 / 32	24 35 33 18 / 3 16	26 3 24 19 2 / 9	32 19 3 2 5 / 13
12 35 8 4 13 / 10 1	1 13 30 2 35 / 4	1 2 35 13 38	2 12 26 28 8 / 6	35 1 12 13 / 24 17 4	13 35 9 1 17 / 7	1 13 3 39 18 / 17 24	2 19 26 35 / 22 36 34	19 13 32 2 / 24
10 15 12 22 / 40 35 28 36	19 5 35 38 9 / 13 3	1 19 3 5 35 / 13 4	10 35 28 38 / 19 15 24	3 14 9 37 1 / 28 13	3 35 5 10 40 / 28 29 18	24 35 3 4 39 / 12 25	35 28 21 36 / 10 3 40 31	1 24 19 31 / 26 17 32
3 9 29 17 26 / 2 16	28 5 2 29 35 / 10 13 27	28 10 5 13 / 35 2	19 28 2 30 / 35 20 34	19 35 4 9 2 / 40 1	40 28 13 2 9 / 35 29	24 2 9 25 5 / 16 39 36 19	24 13 36 2 / 17 35 28 9	35 24 13 17 / 14 4 28 9
10 35 37 23 / 25 26 7 24 4	35 25 10 7 / 19 5 1	35 37 1 10 7 / 5	24 37 10 5 / 25 35 28 1 4	3 40 10 35 2 / 25 5	3 10 35 5 24 / 28 2 17	3 22 13 24 / 28 5 1	35 19 2 36 / 10 13 28 3	25 35 2 13 5 / 24 39 3 4
28 15 19 37 / 3 24 40 30	2 35 38 37 / 24 31 19 4 / 28	35 19 2 24 / 28 13 16 4	19 1 35 24 / 28 16 18 10 / 3	35 37 24 1 / 32 10 3 36 / 30	28 3 24 27 / 15 32 37 1	28 39 2 10 / 30 24 35 22	35 3 28 32 / 24 13 2 16	28 26 24 2 / 13 3 5
24 28 37 2 / 32 35 1 9	24 10 3 28 6 / 37 19 35	6 10 24 5 3 28 / 13 20	3 5 10 24 13 / 28	24 5 2 13 10 / 28 3 32	28 24 2 6 10 / 5 3 32	35 39 2 37 / 13 24 12 1	28 24 6 19 2 / 10 32 3	32 1 35 24 6 / 10 2 31

Improving Feature → / Worsening Feature ↑			24 Function Efficiency	25 Loss of Substance	26 Loss of Time	27 Loss of Energy	28 Loss of Information	29 Noise	30 Harmful Emissions	31 Other Harmful Effects Generated by System	32 Adaptability/ Versatility	33 Compatibility/ Connectability	34 Transability/ Operability/ Controllability
						Efficiency							
Manufacture/Cost	41	Manufacturability	1 10 15 16 3 6 25	19 34 33 9 15 2 12 13	3 4 35 15 20 28	19 35 2 13 10 24 16 15	25 24 16 10 2 22 37 6	24 9 2 39 13 25 5	35 10 5 36 21 24	35 10 5 21 24 39 29 7	1 3 25 13 15 30 19 29	28 3 6 2 13 15	2 5 28 13 16 25 15
	42	Manufacture Precision/Consistency	3 10 14 40 7 13	10 31 35 24 36 37 3	28 15 5 18 32 26	2 16 13 32 35 29 31	13 10 2 34 7 1 24	2 13 7 37 35 9 18	3 10 40 24 19	10 17 35 4 23 34 26 33	35 7 13 1 4 17 12	13 4 9 28 15 2 1	19 3 1 32 35 2 25
	43	Automation	17 3 2 28 19 29 15	10 4 24 5 35 3 12 18	15 28 35 24 29 31 2 5	28 21 3 13 34 24	5 3 28 33 35 24 25	31 14 9 12 24 13 3	1 24 23 35 21 13 3 4	24 2 11 3 23 19 7	28 1 29 10 12 4 14 35	6 13 2 17 10	5 25 21 10 23 17 3 12 1
	44	Productivity	2 13 10 3 29 28 24	35 12 2 34 14 3 24 9 5	10 3 6 24 34 1 7	28 35 15 14 9 5 19 24	10 2 3 24 25 13 4 37	9 14 1 2 31 4 17	35 25 13 2 34 21	24 39 13 35 3 40 5 17 18	28 15 29 35 1 10 17 40 36	6 28 13 35 4 24 32	28 7 26 24 10 1 35 25 15
	45	System Complexity	10 28 2 13 9 35 3 12	35 28 10 13 2 29 2 24	3 25 6 29 24 4 19 10 25	35 28 13 10 2 24 38 34	25 7 6 24 19 32 1	24 9 2 13 3 6 25	5 24 15 2 21 13 35	1 12 19 35 36 21 29 22	29 28 1 24 15 25 37	6 28 13 35 4 24 32	4 26 24 6 25 28
	46	Control Complexity	35 10 24 13 1 6 28 37	35 15 30 13 2 29 34 4	15 3 13 7 2 4 19 10 25	35 15 13 3 19 2 4 25	25 10 7 3 4 26 24 1 32	10 3 7 15 25 9 2 4 26	10 15 1 5 13 39 4 24 38	19 12 16 22 9 28 37 34 27	1 7 28 25 26 22 35 10	6 10 13 1 2 24	2 10 24 25 7 22 1 13
Measurement	47	Ability to Detect/Measure	28 10 2 6 7 24 25	1 10 24 18 35 24 34 31 16	24 28 2 10 6 32 37 26 34	26 10 24 37 35 27 6 1	24 7 25 37 1 6	1 24 2 37 25 7 13	28 2 25 37 13 35 7 24	2 10 34 21 38 6 12 37 7	1 26 13 15 35 19	25 1 13 15 35 28	2 5 13 25 15 35 7
	48	Measurement Precision	28 10 2 6 7 24 25	28 10 13 2 35 24 34 31 16	24 28 2 10 6 32 37 26 34	26 10 24 37 35 27 6 1	24 7 25 37 1 6	35 2 24 9 7 13 39	35 2 24 9 7 13 39	10 3 24 39 22 13 36 16 26	35 2 10 13 24 6 1	2 6 10 15 19 24	25 13 1 15 6 24

40 Inventive Principles

#	Principle	#	Principle
1	Segmentation	21	Skipping
2	Taking Out/Separation	22	'Blessing In Disguise'
3	Local Quality	23	Feedback
4	Asymmetry	24	Intermediary
5	Merging	25	Self-Service
6	Universality	26	Copying
7	'Nested Doll'	27	Cheap Short-Living Objects
8	Anti-Weight	28	Mechanics Substitution
9	Preliminary Anti-Action	29	Pneumatics And Hydraulics
10	Preliminary Action	30	Flexible Shells And Thin Films
11	Beforehand Cushioning	31	Porous Materials
12	Equipotentiality	32	Colour Changes
13	'The Other Way Around'	33	Homogeneity
14	Curvature	34	Discarding And Recovering
15	Dynamization	35	Parameter Changes
16	Partial Or Excessive Actions	36	Phase Transitions
17	Another Dimension	37	Thermal Expansion
18	Mechanical Vibration	38	Strong Oxidants
19	Periodic Action	39	Inert Atmosphere
20	Continuity Of Useful Action	40	Composite Materials

Enhanced Performance

	35 Reliability/ Robustness	36 Reparability	37 Security	38 Safety/ Vulnerability	39 Aesthetics/ Appearance	40 Other Harmful Effects Acting on System
	2 3 35 9 28 27 33	1 36 13 23 25 27	1 24 10 13 5	6 35 28 9 1 24	30 24 16 32 3	35 39 2 29 22 21 19 11 8
	28 25 13 1 5 11	3 10 1 25 30 29	2 10 24 3 25	4 3 31 35 30	2 3 17 32 7	10 28 9 23 2 24 33 35
	12 28 23 7 35 31	13 35 4 2 28 37 17	25 13 28 2 10 7	31 30 16 39 24 40 5	3 16 22 35 9 2	2 25 1 23 33 13 24 29 3
	3 1 35 10 14 24 39 9	1 25 9 2 13 17 24	10 24 1 28 7	10 39 1 18 31 24	2 13 22 1 17 7	35 13 24 33 40 2 11 14
	35 1 13 40 2 33 6	2 3 13 28 1 27 2	28 5 24 10 19 13	24 10 26 35 4 13 14	5 32 35 22 24 33 3	40 19 15 39
	23 3 1 2 10 35	13 25 1 7 10 24 28 35	24 2 13 10 7 4 26 5	10 25 11 39 2 7 24 37 35	7 10 37 5 13 2 25	1 1 29 27 9 12 37 5 2 39
	28 1 40 26 35 2 8 10	26 2 5 12 10 27 1	1 10 3 32 23 13	24 2 10 26 9 12 13	2 28 26 24 13 7 31 3	19 28 22 3 30 29 4 9
	24 5 23 13 11 25 35 1	28 24 13 32 1 2 11	13 27 24 2 6 11	12 24 9 23 13 10 28	28 35 7 31 13 24	28 24 26 22 2 13 35 39 29

Improving Feature		Worsening Feature							
		Manufacture/Cost						**Measurement**	
		41 Manufacturability	42 Manufacture Precision/ Consistency	43 Automation	44 Productivity	45 System Complexity	46 Control Complexity	47 Ability to Detect/ Measure	48 Measurement Precision
Manufacture/Cost	41 Manufacturability		3 16 25 12 24	1 10 28 8 13 25	1 13 15 14 5	27 26 1 5 10 9	25 3 19 13 6 32 39	6 28 1 13 11 10	35 12 1 6 28 13 15 29
	42 Manufacture Precision/ Consistency	25 13 24 15 19 26 28		25 15 24 13 17 28 26	2 10 39 18 32 16 5 26	2 16 18 26 4 28	28 25 37 10 26 7	28 26 32 18 9 25	28 32 18 9 3 25
	43 Automation	1 13 12 26 21 10	25 28 26 3 17 12 18 4		12 26 2 5 35 15 10	15 24 28 10 13 4 17	28 3 4 17 37	25 17 1 27 28	13 24 28 25 10 32
	44 Productivity	1 10 24 35 28 6 19	3 32 26 18 1 25 35	10 5 12 26 35 4 13		6 12 10 1 28 27 17 31	25 1 19 7 24 16	4 32 37 25 28 18 35 24 13	1 28 19 37 4 13 24 3
	45 System Complexity	3 13 1 28 26 35 10 6	2 24 13 3 35 26 32	28 3 1 24 10 13 17	26 12 29 8 17 1		3 37 25 26 7 28	28 10 32 37 15 24	28 26 10 2 34 7 37
	46 Control Complexity	13 2 35 10 28 21	25 2 24 13 10 39	1 28 25 10 21 34	35 1 10 21 28 15 13 25	28 15 37 7 2 13 35 29		28 32 37 3 10 6 24	10 26 2 37 7 28 3 4 32
Measurement	47 Ability to Detect/Measure	5 28 37 11 2 13 29 24	2 28 10 25 2 5 13	10 2 28 25 5 26 37 1 21	2 28 10 35 25 5 18 26 37	28 37 10 15 3 24 25 32	28 32 37 3 10 6 24		28 26 32 24 3 13 37 10 18
	48 Measurement Precision	3 25 28 13 35 24 1	28 26 24 23 25 1	28 24 26 2 10 1 3 24	28 26 10 3 13 24	3 35 10 27 1 13 28 26	3 25 10 7 13 1 23 19	26 28 24 10 13 1	

Worsening Feature →
Improving Feature ↑

40 Inventive Principles

1	Segmentation	21	Skipping
2	Taking Out/Separation	22	'Blessing In Disguise'
3	Local Quality	23	Feedback
4	Asymmetry	24	Intermediary
5	Merging	25	Self-Service
6	Universality	26	Copying
7	'Nested Doll'	27	Cheap Short-Living Objects
8	Anti-Weight	28	Mechanics Substitution
9	Preliminary Anti-Action	29	Pneumatics And Hydraulics
10	Preliminary Action	30	Flexible Shells And Thin Films
11	Beforehand Cushioning	31	Porous Materials
12	Equipotentiality	32	Colour Changes
13	'The Other Way Around'	33	Homogeneity
14	Curvature	34	Discarding And Recovering
15	Dynamization	35	Parameter Changes
16	Partial Or Excessive Actions	36	Phase Transitions
17	Another Dimension	37	Thermal Expansion
18	Mechanical Vibration	38	Strong Oxidants
19	Periodic Action	39	Inert Atmosphere
20	Continuity Of Useful Action	40	Composite Materials

Chapter 11
TRIZ Separation Principles

Frequently, there is an underlying physical property which is the root cause of the contradiction between the two different parameters as illustrated in the contradiction table. For example, let's consider the contradiction of wanting to improve "speed" (parameter #14) and seeing a resulting "loss of energy" (parameter #27). There are some very good suggested inventive principles at this intersection such as principle #19, "periodic action" (do we need the speed all the time?), principle #35, "parameter change" (can we store some of the excess energy in a phase change material?), or principle #1"segmentation" (do we need the required speed at all points in the process or machine?), and finally principle #14, "curvature" (can we redesign the body to reduce wind loss?). We need to ask the question, "What is the underlying physical conflict that is causing this loss of power? It might be friction. We want friction to be there at certain times or under certain conditions, but not at other times or under other conditions.

In TRIZ, we call this type of contradiction a *physical contradiction* and it is frequently the root cause of the conflict *between* parameters. It sometimes takes more thinking time to think about this and identify a specific physical property that is at the root cause of the conflict. This can't always be done, but when it is possible, it sharpens the contradiction and makes the problem easier to solve.

This can be pictured as follows (Fig. 11.1).

In this view, a parameter of a system needs to have two different properties for different reasons. For instance we want something to be thick for one reason and thin for another. Think about a fuse in this context. We may want something to be heavy for one reason and light for another. Think about this contradiction in the context of a car's weight for comfort and safety versus fuel economy. This type of analysis leads us to a much smaller number of inventive principles, under which the 40 principles can be grouped. We'll discuss this later in the chapter.

Let's look at a simple wrench (Fig. 11.2).

You have used them for years and they haven't changed much. Some novelty has also been introduced in the wrench head to allow it to grab more securely worn and smooth bolt heads. Their metal and metal surface may have changed slightly, and they have become adjustable (this had made the wrench more *dynamic* and *ideal*,

J. Hipple, *The Ideal Result: What It Is and How to Achieve It*,
DOI 10.1007/978-1-4614-3707-9_11, © Springer Science+Business Media New York 2012

Fig. 11.1 A physical
contradiction

Parameter #1 vs. Parameter#2

(controlled by)

Physical Property #3

Fig. 11.2 A simple wrench

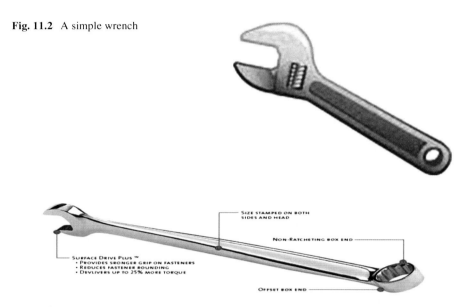

SIZE STAMPED ON BOTH
SIDES AND HEAD

NON-RATCHETING BOX END

SURFACE DRIVE PLUS ™
· PROVIDES SRONGER GRIP ON FASTENERS
· REDUCES FASTENER ROUNDING
· DEVLIVERS UP TO 25% MORE TORQUE

OFFSET BOX END

Fig. 11.3 The X-Beam® wrench

replacing many different wrenches). But what's still a problem that has been around since the wrench was invented? When the wrench is in position to grab the head of bolt, the conventional design requires that the wrench head must be parallel to it. That means the wrench is also perpendicular to your hand which has to supply the pressure in the form of force and leverage. The narrow edge of the wrench handle (not the flat side which is parallel to the bolt head) is now perpendicular to your hand and cuts into the palm of your hand. We've put up with this painful geometry for decades. Let's think about how a wrench is made—casting molten metal into a mold. This mold is used thousands of time, so once we make it, there's no future new capital cost. Maybe the fact that one of these molds was made in the current shape a long time ago, has prevented the development of this "new" X-Beam™ wrench which received the Arthritis Foundation award in 2006 (Fig. 11.3).

Why did the wrench handle *ever* have to be uniform? (Recall the symmetry bias we have). Who said? Did we spend any more money on this type of mold? I doubt it. Just thinking about changing the mold (an existing, not new) resource makes the product more *ideal.* This simple example illustrates one of these higher level inventive principles—*separation in space.* The wrench is one geometrical shape at one point and a different shape at another.

Fig. 11.4 Restroom entrance without a door

Table 11.1 Principles applicable to separation in space

1. Segmentation	2. Taking out/trimming	3. Local quality	4. Asymmetry
7. Nested doll	13. Other way around (do it in reverse)	14. Curvature	17. Another dimension
24. Intermediary	26. Copying	30. Flexible shells/ thin films	

Have you ever thought about how this same principle is used in most airport restroom entries? Luggage is sometimes cumbersome and opening doors is a hassle. Why not just change the geometry of the entrance to get the door's *function* without its existence and use a simple barrier (separation in space) (Fig. 11.4).

Think back to the Boeing jet engine cowling example we discussed earlier. We could have framed this as a physical contradiction instead of a two parameter contradiction by seeing that there was actually an underlying physical property that was in conflict with itself and that is the diameter of the engine cowling. We want it to be long for one reason and short for another. By making the diameter nonuniform (separating in space), we resolved the contradiction.

Mann has grouped the 40 principles [1] around these broader categories and these are the principles most applicable to "separation in space" (Table 11.1).

Have any of you held a bottle's metal lid under hot water for a few seconds to make removal of the cap easier? What are you doing? You are taking advantage of the fact that glass and metal have different coefficients of expansion and the metal will increase in volume faster than the glass, allowing easier removal. You have separated the cap and glass in space through the use of the difference in a relevant physical property. You have also used the "another dimension" problem solving principle as metal increases in volume far faster than glass.

You have seen this separation principle used in highway design with lanes to maintain separations and in airline pricing strategies. Separation in time and space *both* are used by airlines to segregate fare charges for passengers—where you sit and when you buy your ticket. We see this same use of separation principles in admission lines in highly visited amusement parks. We have also seen this used in the allowing of nonviolent criminals to buy their way into "nicer" jail accommodations in California (http://www.digitaljournal.com/article/173864/Upgrade_Your_Jail_Cell_For_80_Bucks_A_Day).

The three other separation principles can be described as *separation in time, separation between parts and the whole, and separation upon condition.*

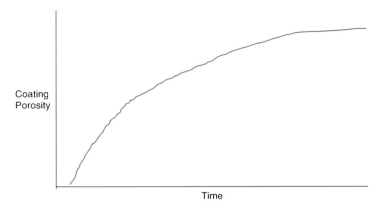

Fig. 11.5 Coating porosity versus time

Let's look at each one in turn. First, separation in time. When we run a production operation, we frequently want everything to run consistently and without variation so that uniform product is produced. In fact, we use tools such as Six Sigma to ensure there is *no* deviation. Many times this is a correct strategy, but let's considers the following case:

An electroplating process is being run that requires an impervious, high density surface. Because of this, the plating rate is very slow to avoid porosity developing from the encapsulating trapped gas from the solution. It is desired to increase the capacity of this plating operation without building a new plating line.

How would we approach this problem as a Six Sigma Black Belt? We'd probably set up a process to collect a lot of data from the plant to see how fast we could push the process until void volume in the coating increased to the point where it did not satisfy customer requirements. This would no doubt involve a lot of interaction with the customer investigating the various coating levels and the effect on the final products' performance. Then we would settle on a process control system to maintain that level of coating porosity. The graph of the final result of this work would probably look something like the graph shown in Fig. 11.5.

How might we approach this problem from a TRIZ perspective? Let's go back and slow down and start with step one in our mental problem-solving algorithm: the Ideal Result. If we go back and read the description of the problem very carefully, it says we want an impervious *surface* not an impervious *coating*. Now I am not trying to trick you here—I am trying to illustrate with a real case what happens to us all the time in problem solving. We too often do not spend sufficient time *defining* the problem. This is what TRIZ forces us to do! The first step in this process is to define the Ideal Result. What is important in this case is that the *surface* is impervious. The characteristic of the coating underneath the surface may be important for other reasons, but if the surface is impervious and does not allow transfer of liquid or gas, then its characteristics can be different than the bulk of the coating. If we identify surface

Table 11.2 Principles applicable to separation in time

9. Preliminary anti-action	10. Preliminary action	11. Beforehand cushioning
15. Dynamism	16. Partial or excessive action	18. Mechanical vibration
19. Periodic action	20. Continuity of useful action	21. Skipping
26. Copying	34. Discarding and recovering	37. Thermal expansion

porosity as the key, how does that change our thinking about how to solve the problem? We could run the coating process very fast at the beginning and possibly up to as much of 90 % (or more!) of its entire time cycle and then slow the process down at the very end to produce an impermeable surface. The overall productivity will increase dramatically. Uniformity is not necessarily innovation or optimization! Once we have this new concept defined, we can set up a series of experiments to slow the process down at various stages, measure the barrier properties of the coating, and when we have found the ideal place to switch from fast to slow deposition, design a control system to produce Six Sigma results for this *nonuniform*, but *more ideal*, coating. This problem illustrates profoundly why TRIZ problem solving places so much emphasis on problem definition and also why TRIZ use is frustrating to many individuals who want to jump into the solution space prior to complete and thorough problem definition.

If we look at the TRIZ 40 principles, it is possible to group some of them under the "separation in time" category [1] (Table 11.2).

Let's think about some of the many ways we have seen "separation in time" used. In an automobile, we have been using timing cylinders for decades. We plan our bill payments to come right after our paychecks arrive. We plan production runs months in advance to match up with seasonal product demands that coincide with weather patterns or holiday purchases. Recently, Intel has announced a product which does not use its entire power supply unless it is needed.

In chemical processing, we try to minimize potential hazards by minimizing the time spent on especially hazardous reactions ("skipping" if possible!). In shipping, scheduling, and transport, we try to use "just in time" concepts to minimize the time a product sits in inventory providing no value. We use vibration (separation in time of an applied force) to loosen frozen or stuck objects.

Are airline fares constant with time? No, they are changed constantly with time as a function of competitor activities and capacity fill of the particular flight. In general, they head higher as the departure date approaches, but in the case of a 1/3 full plane that is going to fly anyway, the fares might be dramatically reduced as departure time approaches. Both approaches are using separation in time, but in different directions.

Think about how "preliminary anti-action" is used in business. To avoid a potential hostile takeover, "poison pills" are embedded into corporate structures that trigger costly penalties should an unwelcome buyer attempt to purchase a company. In that same vein, "beforehand cushioning" is used to provide "golden parachutes" to executives who might be out placed in an unfriendly acquisition.

Table 11.3 Principles applicable to separation between parts and whole

1. Segmentation	5. Merging	6. Universality
8. Counter-weight	12. Equipotentiality	13. "Other Way Around"/ Do it in reverse
22. Blessing in disguise	23. Feedback	25. Self-service
27. Cheap, short living objects		

Let's now look at *separation between parts and whole or system/sub-system/super-system*. In this category, Mann lists the following 40 principles [1] (Table 11.3).

How many devices and instruments have you seen that are segmented to allow flexibility? Medical scopes and drain cleaners come immediately to mind. How many devices do you own or use that have disposable elements (cheap, short living objects)? Batteries are a classical example here where the supersystem (the camera, the flashlight, or the computer) stays intact, but the power supply is replaced. Your car and other sophisticated machinery which are expensive to maintain all have internal diagnostic systems providing "feedback." The internal process control loops in a chemical plant are another example where we use this principle. The output from these control loops frequently provide input to external control loops.

Do you remember what you were told in high school chemistry class about adding water to sulfuric acid? Don't do it! The heat of solution would cause the water/acid mix to "spit" out and possibly burn you. Add the acid to the water ("Do it in reverse") so that the much larger heat capacity of the water can absorb the heat. You would also find this principle as one suggested in looking at the technical contradiction of "safety" (parameter #38) versus possibly 15 worsening parameters, including weight, area, volume, shape, amount of substance, duration of action, speed (a speed limiter on a car throttle or a restriction on quantity addition rate in a chemical reaction?), force/torque, function efficiency, loss of substance, and others. Pre-placement of goods in assembly lines solves the contradiction of control complexity versus connectability.

Counterweights are used frequently to maintain balance in an overall system when individual components shift position.

Think about "feedback" in this context. How often have you visited a restaurant, and the service wasn't quite up to par (but OK) and you complained, left less of a tip, and were given a discount coupon for your next visit? Your one visit was within the "supersystem" of many visits and the restaurant's total business. Many customer surveys utilize this way of thinking to find out about the quality of their service as well as what else you might want to buy that would expand their business.

In thinking about "segmentation", a recent interesting business example of this principle is Dow Corning's split off of a separate organization called Xiameter® which supplies only certain large volume products to customers who do not require technical service. They receive special discounts, but no support if they need it.

Finally, let's consider *separation upon condition*. This is where a physical property of a system changes in response to an external condition. Think about anti-lock brakes as an example we see in the auto industry to overcome the natural temptation of a driver to push harder on the brake pedal when sliding on ice, when what is required is repeated pumping of the brakes. The anti-lock system senses the

Table 11.4 Principles applicable to separation upon condition

28. Mechanics substitution	29. Pneumatics and hydraulics	31. Porous materials
32. Color changes	35. Parameter changes	36. Phase transition
38. Strong oxidants		

slippage and automatically backs off the brake pressure no matter what the driver is doing.

Here are the 40 principles most often used when dealing with separation upon condition situations, as listed by Mann [1] (Table 11.4).

Notice that many of these involve a physical change to a substance as a result of external changes. Look at all the areas we use these principles to indicate a *change in condition*.

We cause traffic to stop and go by the change in a light color from red to green. Some frozen food products have indicators which change color if they have been thawed to alert consumers that the food has not been kept frozen throughout the supply chain. Phase change (liquid to solid, and liquid to gas) is used to absorb large quantities of heat. Porous materials can allow passage of only certain size materials or molecules. This is the basis for all filters, hazardous emission control, and particulate control in high tech manufacturing of printed circuit boards. It is the basis for reverse osmosis to produce clean drinking water from salt water or industrially contaminated water streams.

We see the application of pneumatics and hydraulics to buffer a moving object (a car) against bumps or holes. We use mediators in labor disputes to separate the arguing parties within the super-system of a labor negotiation.

When we think about physical properties and geometrical shapes, we often seem to have "uniformity genes."

Let's consider a simple illustration of how one might use all of the separation principles in a problem situation that you can easily relate to. Consider the challenge of a highway design engineer who is confronted with the problem of two highways that are supposed to connect with each other for the first time. What would you say is one of the fundamental technical contradictions? There may be many subtle ones such as scenery or environmental disturbance, and there may be some more practical ones such as cost and construction materials, but certainly a primary one would be that no cars on one highway collide with cars on the other highway. This would supersede any other concern. Let's look at each separation principle in turn as applied to this problem.

Separation in space: parallel roadways, yield lanes, upper and lower decks
Separation in time: stop lights, timed lights, sequenced lights for on ramps
Separation between parts and whole: roundabouts
Separation upon condition: yield lanes

Are there others you might add?

As you can see, both the TRIZ 40 principles and separation principles are very robust and useful in virtually any situation where there is an apparent conflict between system requirements and system capability. Usually we find out that compromising and uniformity is not the pathway to breakthrough solutions.

Business Case Study

Let's look at some additional applications of these principles in the business arena.

A company may be studying the acquisition of another organization or intellectual property. For a variety of reasons, including many that are legal and antitrust oriented, these discussions must be kept totally separate from on-going business decisions. Frequently this problem is resolved by actually separating the acquisition team physically from their normal office space to minimize the potential of information contamination in either direction. This example illustrates the use of the "separation in space" principle in a business and organizational sense.

In this current day and age of the Internet and web-based communication, we see the "do it in reverse" principle used in marketing and sales. Customers, who have a need, type a search word into a browser, and find a supplier as opposed to the supplier sending out a salesperson to call on customers. We also see the "nested doll" principles being applied at the same time. If we want to find the least costly insurance policy, for example, we type in our information and within hours, several insurance companies will provide a quote without ever having to have spent any money on sales travel.

A large multinational corporation is frequently organized in a "nested doll" fashion. Within its structure, there are numerous separate businesses that are focused on particular industries and industry segments. In fact, in some organizations, these may even be individually named with no apparent connection with the parent company. An example of such a company headquartered in the USA would be ITW. Though also known as Illinois Tool Works, a search of its web site will show not only at least eight differing industry segments but just in the case of one of these, packaging, there are eight different companies supplying various needs in this area. In most cases, they have chosen to maintain the identity of the acquisitions they have made to supply customer needs in these various areas. Siemens, a global corporation based in Germany, also uses nesting in the sense of many different business units under one umbrella, but chooses to link the corporate name, Siemens, with all of them. A different approach, but the same strategy.

Let's also take a look at the application of these principles to an organizational problem that might be typical in a large organization. This case is an actual one presented by a large group of medium-sized companies in a medium-sized metropolitan area. This group of companies characterized their business climate as follows:

1. Everyone is "overloaded"
2. Plates are full and getting fuller
3. The world is full of miracle tools
4. Objective sources to evaluate, compare, and assess appropriate application are few and far between

In this context, these companies described what they called the "paradoxes" of innovation. (If you look in a thesaurus, you'll see that a paradox is just another word for *contradiction*). Here is how the group described these paradoxes:

1. Innovation needed to be *"someone's job" and yet, at the same time, "everyone's job."* The group interpretation of this contradiction was the struggle to have much of the organization focused on today's customers, specifications, billing and invoicing, and other short term activities, but at the same time recognizing that innovation was also a key organizational requirement. Today's products and services will eventually be replaced by new ones or new delivery systems. If everyone was focused on innovation, none of today's required business requirements would be fulfilled; on the other hand, if no one was worried about innovation, the business and company would ultimately fail for not having planned for the future.

2. Concern about *"inside" business focus versus "outside" business focus*. What was discussed here was the conflict (contradiction!) between being focused on your current products, customers, and business and the longer-term vision of what that might be. For example, consider the example of the paint pan supplier mentioned in the previous discussion about the Stanley Black & Decker Pivoting RapidRoller™. What percentage of your resources should be spent on reducing cost, optimizing geometry, choosing metal composition, and distribution economics vs. the market's long-term vision of whatever "holding paint" might mean. Talking to existing customers versus potential customers may give very different answers to this type of question. How should this be balanced? What if the answer to the question led you in a completely different direction than paint pans? Would you go there? What would be involved?

3. *Chaos versus discipline* was the next item of concern expressed by the group. What was meant by this? This was recognition that frequently, in the initial stages of a new product or new business, things are very chaotic. It may not be clear what the target is, how to get there, who and what the competition may be, what costs might be, and what all the issues may be in the intellectual property arena. However, as the innovation moves toward commercialization and commitment of large sums of money and people resources, the process must become more disciplined. How do we balance these two extremes? When do we make the transition? What determines this balance?

4. *Passion versus objectivity* was another paradox (contradiction) cited by the group. This has an overlap with the previous one in that early on in a new product's development; it is frequently the passion of the inventor that drives the development and evaluation of the concept. At some point in time, however, the economics and other practical aspects of implementation and commercialization become very important and passion cannot overwhelm facts at some point in time. How do we blend these two important aspects? How do we balance them?

5. Last, but not least, there was concern about the challenge of *balancing risk and job security*. This issue can be a very serious one, especially in large

organizations with large numbers of people dedicated to profit making existing businesses. Think about your own organization and situation. What happened the last time a significant major new product introduction failed? What happened to the people involved? Do they still have jobs? If they do not, what kind of message did that send to the rest of the organization? Now we are not talking here about the situation that could exist if the new product or business development team made some very stupid decisions and need to suffer the consequences. We are talking about the case where senior management, despite all "green" go signals, arbitrarily decided to not pursue a new business for reasons or circumstances outside the control of the product team. How many volunteers will raise their hands the next time?

As an exercise, think about these five "contradictions" and apply the TRIZ separation principles previously discussed to each of them and come up with at least two ideas. The actual group responses are in the appendix and you can compare your answers with those later on.

Innovation contradiction	Resolution idea suggested by separation in
Someone's job versus everyone's job	Time: _____
	Space: _____
	Parts and whole: _____
	Upon condition: _____
Inside business versus outside business focus	Time: _____
	Space: _____
	Parts and whole: _____
	Upon condition: _____
Passion versus Objectivity	Time: _____
	Space: _____
	Parts and whole: _____
	Upon condition: _____
Chaos versus discipline	Time: _____
	Space: _____
	Parts and whole: _____
	Upon condition: _____

Innovation contradiction	Resolution idea suggested by separation in
Risk versus job security	Time: _____
	Space: _____
	Parts and whole: _____
	Upon condition: _____

To summarize this discussion about contradictions:

1. The resolution of contradictions (as opposed to minimizing the pain) is the key to breakthrough inventions.
2. Contradictions can be of two types—technical (between two or more different parameters in a product, system, or service) and physical (one parameter whose desired properties are in conflict with each other).
3. How contradictions are resolved provides the basis for the TRIZ 40 inventive principles as well as the separation principles.
4. Focusing on contradictions within the design or operation of a product, service, or system is the primary way of achieving a breakthrough in the design or operation of the product, service, or system.

Exercises

1. For each of your current products, services, or systems—what is the contradiction that keeps you from improving its design or operation? How would you characterize it in the framework of the TRIZ contradiction table? What ideas can you develop from the inventive principles suggested by the principles highlighted at the intersection of these conflicting parameters?
2. When you look at your current products and services, do you see examples where a single parameter is in conflict with itself? Examples might include weight, space requirements, area, number of people, geographic coverage, time spent in a certain area, or delivery time or mechanism. If so, take each of the separation principles (time, space, parts/whole, upon condition) and generate at least one idea for resolving the contradiction.
3. For a product or service that you currently perceive as having no major problems or issues, use the separation principles to change the design or operation in terms of space, time, parts of the system, or condition and come up with at least one new version of the product or service that may provide a new business opportunity.
4. Think about the last time you had a brainstorming or idea generation session and you heard the words "that's a good idea, but…" What did you do then? Discard the idea? Why? If you can remember the details, reframe it in terms of a

contradiction, get out the table (or use the separation principles) and resolve the contradiction(s).
5. Think about the next generation product or service you have on the drawing board. If you are using a tool or process such as Design for Six Sigma, how could you use the tools of TRIZ to head off future product or service issues?

Reference

1. Mann D (2010) Hands on systematic innovation. Lazarus Press, UK, p233

Part III
TRIZ Strategy
and Analytical Tools

In this section, we will review how some of the previous TRIZ tools can be used in a strategic as opposed to a problem solving fashion, as well as TRIZ tools for analysis and strategy. These aspects of TRIZ developed over time as the analysis of inventions provided additional insights into the predictive science of technology development.

Several of the previously discussed concepts can, in a general sense, also be used in a strategic sense as opposed to a problem solving sense. For example, a continuous effort to make a product, system, or service more "ideal" (even if no one is asking for an improvement) needs to be part of strategic thinking. Using a TRIZ-based resource analysis coupled with the concept of trimming can be a powerful combination in long-term cost reduction efforts.

In the rest of this section we will review more specific TRIZ analytical tools, again derived from the study of technology evolution across many fields, focused on strategy and new product and new business development.

Chapter 12
Lines of Product System Evolution

In the study of the patent literature, we not only see common trends in the repeated use of the same general inventive principles, but also common trends in how products and systems evolve. These "lines of evolution" allow us to qualitatively, forecast how products and businesses will evolve. We need to emphasize that these lines are qualitative not quantitative. These lines do not tell us in what time frame these changes will occur. Another aspect of these lines is that they contain many discontinuities that are not natural extensions of current products and businesses. This fact forces organizations to consider how they will handle these discontinuities. This may require substantial changes in the type of people hired, the schools where recruiting is done, and what groups and societies are of interest and are supported. This can be a difficult challenge for many organizations.

There are a minimum of eight of these lines and these can be dissected into many more. Some of them overlap significantly with concepts already discussed, such as "systems and products evolve to a more ideal state over time" and "systems evolve through the resolution of contradictions." Since we have already covered these concepts, we will focus on those lines which have not been discussed in some form and are not intuitively obvious from previously presented material. The lines we will review, discuss, and present examples for are:

1. Systems and products evolve to become more dynamic and responsive
2. Systems and products oscillate between added useful complexity and simpler design
3. Parts of a system evolve at different rates that may be inconsistent
4. Products and systems evolve through the use of matching and mismatching components and properties
5. Products and systems become integrated into their "supersystems"
6. Products and systems evolve through the use of higher level fields

J. Hipple, *The Ideal Result: What It Is and How to Achieve It*,
DOI 10.1007/978-1-4614-3707-9_12, © Springer Science+Business Media New York 2012

Products, Systems, and Services Become More Dynamic and Responsive Over Time

What does it mean for something to become more "dynamic" or responsive? It means that the product, service, or system reacts in a positive way to external factors that change with time or condition—preferably in a way that is valuable to either the customer or supplier and hopefully to both. Let's consider an example that everyone is familiar with and that we have discussed already—airline seat pricing. The price of a seat on an airline can change dozens of time a day dependent upon:

1. The load factor of a particular flight
2. Fuel costs
3. Competitors' pricing
4. Competitors' flight capacity
5. Number of "free" seats given out through frequent flier programs
6. Time of the year (i.e., holiday seasons)
7. Labor costs
8. Baggage fee and commercial cargo income
9. Time before "take-off"
10. Time of year

Once the plane takes off, an empty seat provides no income to the airline, so as the time for departure approaches and the plane is not full, there can be a last minute sudden drop in pricing. If the plane fills up early and competitors' flights are full, there will be a tendency to keep the fare high up until the last minute. An airline can also decide the amount and quality of food served based on the average fare paid on a particular flight leg. Most airlines are now charging for checked bags. However, if a flier uses an airline's preferred credit card, these fees may be waived. Any feature of a product or system can be reviewed from the aspect of making it more dynamic. Every feature or property needs to be reviewed from this perspective as well as considering all the possible external variables to which a product or system might be made responsive. These same thoughts and ideas can be considered for other services such as hotels, maintenance services, and restaurants.

Where else do we see the recent use of dynamism in product and system design? Intel's new chips use power only to the degree needed at a particular time is one example. As mentioned earlier, General Motors' recent new engine that changes the number of cylinders used based on the required horsepower needed is another. The changing of frequent flier miles as a function of destination, not just miles flown, is another that we see when an airline is trying to promote a new city being served or a city with a new competitor. Cars that incorporate features such as windshield wiper speed varying automatically as a function of car speed are another example. Cars that incorporate a feature that automatically increases radio volume as a function of car speed is another.

Many Web- and Computer-based learning systems automatically adjust to the rate at which a student is learning, as measured by intermediate tests.

Earlier, we showed a picture of a wrench which had been redesigned to alleviate hand pain through the use of separation principles. Let's consider a wrench from the standpoint of this line of evolution. A long time ago, we only had wrenches for a particular size nut. Now we have adjustable wrenches that can be modified through a simple adjustment to handle many different nut sizes. In what ways could such a common tool be made even more dynamic? Responsive (to what)? Could the wrench opening automatically adjust to a given nut size? Could its grip strength be automatically adjusted to the user's hand size? Hand strength?

Other examples to illustrate this line of evolution:

1. Design of antiperspirants whose effectiveness is increased by the release of body moisture
2. Road tolls that change as a function of time of day or traffic volume
3. Slow release fertilizers
4. Bonus payments as a function of company profits
5. Car insurance premiums changing with driving record, car location
6. Flower pricing as a function of certain holidays
7. Cosmetics whose function and behavior changes with skin type and condition
8. Furniture whose shape and style can be changed as a function of a given situation
9. Flexible piano keyboards
10. Adjustable table saws

Several insurance companies now offer reduced rates if the driver is willing to put a data chip in their cars that records driving behavior (http://www.progressive.com/auto/snapshot-discount.aspx). As another example, why should you pay for theft insurance if the car is sitting in the garage?

How could your product, system, or service be made more dynamic?

Feature or aspect	Responsive to what?

For each of the entries you have made, what is the value of the dynamic change? To whom? How could you take advantage of it?

Exercises

1. You have relied on outside contractors to handle variable workloads, but now there is a corporate edict that prevents use of outside contractors. How can you make your group more flexible and dynamic to do the jobs that you believe need done?
2. Your local utility is charging uniform rates for power. Your production operation runs 24 h a day. What can you and the utility do together to save money for both of you?

3. You are responsible for a spray painting operation. Through the use of overspray techniques, you coat every part correctly and uniformly. How can you minimize the loss of materials in overspray using the concepts of dynamism and use of resources?

4. Think about the car that you now own. You're getting older and are not as agile as you used to be. How would you redesign parts of the car to be more responsive to your situation without making things less ideal for other users?

5. Look around you right now and identify every object that is rigid. If these objects were made flexible, what benefits would there be? To whom?

> As you think about making your product, service, or system more dynamic or responsive, don't forget the previous points regarding contradictions and resources. If resolving a contradiction is a key to making a product more dynamic, then use the contradiction table and separation principles to resolve the contradiction. If you can't identify the resources you need, go back and look at the list of resources in TRIZ to rethink what resources you have and how they might be used to accomplish the goal of a more dynamic system or product. Use the TRIZ tool set in a "back and forth" way so that they can feed on each other. Learn to keep track, each time you do a TRIZ exercise, of resources and contradictions. They will provide the feedstock for future thinking, planning, and new product development. As you resolve a contradiction, ask yourself, what is the next one? And remember it might be somewhere else in the system in which you are operating.

Oscillation Between Simplicity and Complexity

We showed a graph earlier (Fig. 7.1) when we were discussing adding useful complexity and trimming as tools for making a system, product, or service more ideal. However, these concepts are also involved in taking a longer term look at how systems, product, and services evolve.

Consider the simple example of lawn care services. In many locations, the need for lawn care fertilization and pest control decreases in the winter time. This means that the company providing these services frequently needs to lay off employees on a seasonal basis. These employees know how to service homes and relate to homeowner needs. We see several of these types of companies getting into the pest control business, which allows them to use their employee resources year round, and allow the homeowner to deal with only one company instead of two. This is a case where the supplier is adding useful complexity and the purchaser of the service may see this as a more idea service offering. Terminix®, a pest control company, is not only offering to provide lawn care, but also pest protection embedded into insulation it will install in your house!

Now think about remote controls for your TV and associated devices. What started out as a simple device that turned a TV on and off and change channels has now evolved into a device that does more functions than most users can use or even understand and which cannot function without at least one other remote control device and the use of a multi-page instruction manual. This is a case where intelligent trimming is appropriate and necessary and this line of evolution would predict will happen. This could come in a number of different ways including "hiding" less used functions, a device that learns from its user and changes its viewing surface, or a more universal remote that can recognize extra devices and be able to change its behavior based on the nature of the total system. When a system, product, or service begins to look like or functions like a "Rube Goldberg" device, TRIZ thinking can safely predict that trimming and simplification will occur. Conversely, if a system, product, or service is not providing sufficient utility (complexity), then useful complexity will be added. Where is your product, service, or system? This needs some serious thought and cannot be answered without some serious thought and analysis. Just because we can add functionality does not mean it is the right thing to do from a business perspective. Similarly, just because we know how to eliminate a function does not mean the user or consumer won't miss it. TRIZ does not provide a direct answer to the question, but the analysis of innovation and patents shows that systems, products, and services oscillate between these two extremes. Look at your product, system, or service from both extremes and honestly assess which direction to focus your innovation efforts.

Make a list:

Useful complexity that can be added

_____
_____
_____

For each of these, ask yourself these questions:

1. Who will buy this added functionality?
2. What will they pay?
3. How do you know?
4. What will this added functionality displace?

If you are not comfortable with either the answers or the quality of the answers, think carefully about proceeding along this path. If you proceed, make sure your assumptions are well documented and tracked as the new product, service, or system is launched.

Now let's look again at the other side: Trimming and simplification. If you know or think you know that your product, service, or system is too complex, where or how can you simplify or trim? How much would cost be reduced? What new market segments would be reached?

Part or aspect to be trimmed	Impact

Exercises

1. Look at every product or service that you provide and ask yourself, for each one, how you could add one more degree of (useful) complexity or one less degree of complexity. What does the new product or service look like? Who would pay for the more complex product or service? Would it displace some other product or service? If the product or service is simpler, what new markets might open up with a possible price reduction? Could a simpler, but elegantly designed product actually be sold for a higher price? (Think about some of the products from Apple).
2. Can you market both a simple and complex version of the same product or service? There are many web-based examples of one level of service being free and a more enhanced version requiring a fee.
3. Look at your competitor's offerings. What would happen if you added one more useful feature or took one away? What opportunities might that present?

Subsystem Parts Evolve at Different Rates

This line of evolution simply states that parts of a system (subsystems) do not necessarily evolve at the same rate. As the car evolved in terms of engine power and speed, these subsystems were added. When first invented and commercialized, its speed was minimal and no one worried about braking, steering, or other systems that we now take for granted. These came along as the automobile engine developed more horsepower which allowed higher speeds and the need for braking and steering capabilities. We also added shock absorbers. Steering became power steering (dynamism, use of resources), brakes became anti-lock (again dynamism, use of resources), and windshield wiper blades became dynamic.

As each of these sub-systems were added, they became limiting under certain conditions, and each was improved using (in hindsight) various TRIZ tools and techniques.

Most products and systems have multiple parts and segments and their evolution does not necessarily occur at the same rate. Each time a product, service, or system progresses along a TRIZ line of evolution or resolves a contradiction, we must immediately look at the entire system and look for the next limiting step or component. It is possible in a system or product that involves a delivery or service component that the service component may be what needs focus and attention as opposed to the "hard" parts of the system. It is also possible that the next limiting system may be external to your business and collaboration with a customer or supplier with this thinking will be of use.

If we think about the evolution of the cell phone, it is part of a supersystem involving towers and satellites in addition to the internal components of the phone itself. At any geographic point, it is possible for the existence of a tower or the capacity of a satellite to limit the transmission of a signal. Increasing the capability of the cell phone itself to do more wonderful things, without considering the capabilities of other rate limiting steps, could be a waste of time and money. How many cases can you think of where the embedded power or capability of a product or software system is used at a small fraction of its capability?

On the business side, consider the case of company or university that has developed a new piece of technology and having no sales or marketing capability in the projected market for the product or service. If this rate limiting step and need is not considered, it is likely that the new invention will not be commercialized to any great extent.

In a chemical reaction system, there is always a "rate limiting" step in a sequence of reactions. Spending money trying to improve the rate of reaction of the most rapid step makes no sense. We need to pay attention to the reaction steps in their reverse order of speed.

In applying this line of evolution to your business or product, it is critical that the entire supersystem it is being used within be considered and that the next limiting step be identified. Then ask which contradictions prevent this limitation from being overcome. What resources are lacking *at this point or step?* Apply the various TRIZ tools we have discussed to overcome this limitation and then look for the next one.

If we look at the evolution of the design of automobiles and bicycles, we can see their evolution in terms of evolving contradictions. When the automobile was first invented, it went only a few miles per hour, so no one really cared about steering and brakes. When the engine power increased, it became important to thinking about stopping the car with more assurance. Steering also became important. As the speed increased, there was also more of a need for shock absorbers. Each of these was a new problem to be solved.

The first bicycle in 1818 had no pedals or transmission. It was simply two wooden wheels, a plank to sit on, and driven by riders pushing with their feet on the ground. Pedals were invented in 1840, 22 years later, and speed increased. Now we needed brakes. Then chains and clutches. At each point, a new contradiction was introduced and had to be resolved with a change in design.

Exercises

1. Draw a diagram or picture of the parts and components of your current product or service. Identify the next limiting part or component, and then the one after that. What is your research or technology plans to overcome these limitations?

How can you use the TRIZ tools to assist you? If the issue is a contradiction, how will you overcome it?

2. Describe or draw a diagram of the supersystem in which your product or service exists or is used. What is the rate limiting step in this supersystem? How can you change your product or service to help overcome this limitation? What would that be worth? To whom?

3. Have you clearly communicated to your suppliers what your rate limiting step is so they can help you?

4. If something you are buying is a rate limiting step, how can you trim it out of your system so that you don't need it anymore?

Matching and Mismatching

This line of evolution says that product and services evolve away from matching to mismatching and then to dynamic mismatching.

Consider audio noise filters. Originally, they were simply noise suppression devices to minimize the influence of outside noise on your ability to listen to an audio recording. Then we had devices which could be manually manipulated for various conditions. Now the most sophisticated head phones constantly monitor external noise and proactively, and constantly, put out an electronic signal exactly $180°$ to the external noise, cancelling out the interfering signal, no matter what its nature.

Copier machines have manual adjustments to compensate for extra light or very dark originals. Dynamic mismatching would provide a way for the copier to sense this need automatically and adjust its controls appropriately.

When we put teams of people together to work on a problem or project, we often deliberately bring in outsiders who may have little direct knowledge of the problem being discussed. This brings a different perspective into the process, often mismatching the preconceived ideas from those who have worked in the area for a long time. Interesting discussions occur and ideas are generated that would have never evolved with a team where everyone thinks the same way or has the same background. There are various psychological assessment instruments (Myers Briggs, HBDI®, Kirton KAI, DiSC®) that can be used to identify the matches and mismatches from many different perspectives.

Even in the simple case of an orchestra, band, or choir, we see dynamic mismatching used all the time to make the music interesting to listen to. Can you imagine how boring a concert would be if everyone sang exactly the same note or played the same note on all the instruments? Composers deliberately mismatch the chords and tones to make the music interesting.

Compensation systems also use this concept frequently to incentivize behavior in changing markets and business conditions. The use of puts and shorts activated as a function of stock prices is another example of dynamic mismatching.

Many machine tools and civil engineering structures have resonant frequencies. If these are approached in the course of normal use, an uncontrolled vibration and oscillation may occur. If these frequencies are known and understood, they can be compensated for.

We usually think it is best to "match" things. It's easier, isn't it? But the study of the patent literature and how systems evolve tells us that is not the preferred way that products and systems evolve. There is nothing in the world that is constant and the more a product or service can adjust to that fact, the more robust it will be and the longer it will survive.

Exercises

1. Make a list of everything in your process, product, or service that is matched in some way (use whatever criteria you want). Now mismatch it in some way (physical attribute, responsiveness, signal nature and type, etc.) What advantages might arise? Challenges? How would the marketing of the new product or service change? Would there be new markets?
2. Go one step further and ask yourself what advantages their might be to dynamically mismatching an input signal or other characteristic of the system. It might be you don't even know how to measure this. What's involved in doing this? What advantages might this new concept provide? New markets? New applications?
3. If you decided to "diversify" a team by bring in different personality styles, how would you do it? Do you have a methodology for doing this? If not, why not? How could this diversity be changed with the current need or situation?

Evolution Along Field Lines

The last TRIZ line of evolution to be highlighted here is evolution along field lines. When we talk about "fields," we are talking about mechanical force, thermal and acoustical effects, chemical fields and reactions, electrical fields, electromagnetic fields, and finally optical fields. When we look at the progression of products, systems, and technology, we see a steady progression along this line. Note, as mentioned before, that the change from one of these line to another is very discontinuous. A chemical expert is unlikely to have substantial knowledge of electronics, electromagnetic, or optical fields. That expert is much more likely to continue to solve a problem or improve a product using the knowledge of chemistry that they have. The same can be said for a mechanical engineer trying to think about new concepts that may involve chemistry.

Let's look at a few basic examples that we can all relate to. First, communication. A long time ago, we sent communication via beating on drums (a mechanical field), then we used smoke signals (a thermal field), followed by the use of a chemical field (writing letters with pencil or ink materials), then we called people on the phone (electronic field), and now we use wireless devices (electromagnetic fields).

Consider cooking and food preparation. We tenderized food with mechanical fields for decades by pounding on them. We then used tenderizers (chemicals). We began the actual cooking process with a thermal field (gas and electric stoves), accelerating and taste changes occasionally by the use of chemical additives. We used high speed convection ovens which combined a mechanical and thermal field. Now we use microwave ovens for many cooking activities. If we look back at the evolution of transportation, we say the use of mechanical fields in the horse and buggy age, followed by thermal and chemical fields in the automobile, and now the use of electronic, electromagnetic and optical fields in advanced batteries, GPS devices, and numerous items inside the car that increase comfort.

In the "picture" or image capturing area, we started out by etching pictures on rocks or in sand (mechanical field), then burning images into rocks (thermal field), then painting and conventional photography (chemical field), and now electronic photography (electronic/electromagnetic fields).

In the chemical process industries, we use high pressure to make some reactions proceed which would not go under normal conditions. Then we discovered, that through the use of heat and higher temperature, an alternative, and sometimes superior way to achieve the same results. Then we discovered chemical catalysts which lowered the temperature required to activate many reactions, eliminating the need for high temperatures and pressure in many situations. And now researchers are looking at ways of applying microwave energy as a potential catalytic route to some reactions. As mentioned before, these changes are very discontinuous, but predictable from a TRIZ perspective.

Another interesting example to think about is how we handle drunk drivers and figure out whether they are "legally" drunk. We first look at their "mechanical" behavior (can they walk a straight line?), we can ask them to say something or repeat what someone else has said (acoustical field), or we can do a breathalyzer test (chemical field). The next step would be an electromagnetic or optical field. This is being developed! (http://www.laserfocusworld.com/articles/print/volume-38/issue-11/features/co2-lasers/optical-nose-technology-competes-with-breathalyzers-and-blood-tests.html).

One important note about these lines. They are not time specific. They do not say when some of these discontinuous changes will occur. However, if you are aware of them, they can stimulate your thinking and you may accelerate their rate of adoption!

Exercises

1. No matter what "field" is the primary driver in your process, what is the next one along the line? Do you have anyone in your organization that has a basic understanding of it and how it operates? Do you have anyone who has ever applied and used it in a practical sense? If the answer is "no," how will you go about obtaining this knowledge?

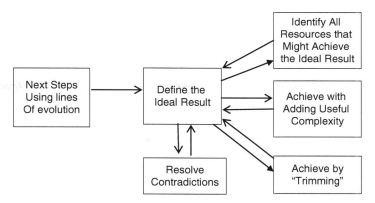

Fig. 12.1 Adding lines of evolution

2. If you analyze the patents in your product or business area over the last 20 years, what kind of trends do you see in terms of field use and evolution? Have you been ahead of the cure or behind the curve in your use of next generation fields?
3. If someone suddenly outlawed the primary fields you use in your manufacturing process, what would you do?
4. If someone suddenly outlawed the use of the primary fields used by your customers in their processes which use your products, how would you help them?
5. If you trace the development of field evolution in your industry, are there "skipped" fields that might provide opportunities? For example, if you see where a product line has moved from a thermal field to an electromagnetic field, could you use a chemical field to some advantage? For example, in the move to microwave ovens, could a cooking additive, sensitive to a particular wavelength of microwave radiation, be added to a food to change its texture, taste, or some other property?

Let's add the use of these lines of evolution to our process diagram (Fig. 12.1).

One point to be made here and that is that the Lines of Evolution can be used independently when there is no apparent problem that we are trying to solve, or a patent that needs improvement or a patent to be circumvented. We can simply take a product, service, or organizational system that we have and ask, "What is the next step along these lines?" Then we can ask to what extent we are working toward that goal, do we have the right internal competencies to pursue, what parallel universes there might be to investigate, etc.

Chapter 13
Combining Upward Integration with Lines of Evolution

We have discussed the use of the various TRIZ tools for specific problems and needs. We have also discussed how the TRIZ lines of evolution can be used to plan and forecast product and system evolution. Now let's combine these thoughts with some of the previously discussed concepts that have strategic as well as product specific uses.

Upward Integration

Let's go beyond the problem-solving aspects of the concepts we've discussed and look in more depth at the business implications and new product and business development. Consider the simple painting case discussed earlier.

You are a supplier of painting tarps. You've been working for years at reducing the cost of your fabric or plastic raw materials. You've been reducing your cost of manufacturing by a combination of outsourcing and modern, computer-controlled process equipment. But over the past several years, you've seen your sales drop despite giving your customers (Lowe's, Home Depot, Sears, Ace Hardware) advance purchase and volume discounts. You have started to investigate and you happen to notice this new Stanley Black & Decker Pivoting RapidRoller™ product and it caused you to think about the implications to your business. It seems like such a simple invention and it minimizes the amount of "slopped" paint as well as the area that needs to be protected. Is there a correlation here? Is it possible that the same kind of thinking might be occurring at the head office of the paint pan manufacturer? Highly likely.

Here's what happening (Table 13.1).

A three part system has been replaced by a two part system and one of those parts has been reduced in size. Upward system integration is usually more *ideal* because it eliminates a part of a system and makes it simpler, both in design and in use, and both of these usually reduce costs as well. There are fewer parts to replace and maintain over time. The history of technology and systems teach us that this upward

J. Hipple, *The Ideal Result: What It Is and How to Achieve It*,
DOI 10.1007/978-1-4614-3707-9_13, © Springer Science+Business Media New York 2012

Table 13.1   Upward integration of painting

↑ Paint in the roller stick and small tarp
Paint in a can plus pan plus large tarp

Fig. 13.1   The Tweel™

integration is one of the constant overriding trends in system evolution. Let's look at some other examples that you can readily recognize.

Automobile audio systems. All of the audio systems that we enjoy are now integrated into our cars as opposed to separate devices that we carry around.

Fax/copying/printing system. Are you old enough to remember having all these capabilities in separate machines that took up a lot of space? Now we have all of these capabilities plus scanning in one machine. If you're a plastics supplier to this industry, you're selling less plastic, and what you are selling has different properties.

Reactive distillation columns. We are starting to see, following significant R&D and process modeling, the integration of chemical reactions internal to a distillation column, potentially eliminating several pieces of process equipment and much capital cost. How does that affect your plans if you are a supplier of process reactors? What kind of instrumentation opportunities might present themselves?

Typing systems. When was the last time you saw a typewriter? Carbon paper? We still have keyboards but they're hooked directly to a screen which can store directly to a hard drive. You can retrieve and Email as many copies as you want without either one. If the suppliers of typewriters and carbon paper had looked "up" in their food chain, and considered some of these possibilities, what might they do? What "business" would they in? Should be in? What acquisitions might they have made to stay in business?

The Tweel™, discussed earlier, illustrates upward system integration with the tire integrated into the wheel superstructure (Fig. 13.1).

The total amount of rubber compounds used in the system have decreased in quantity and no doubt changed in performance requirements. Our diagram looks like this (Table 13.2).

Table 13.2 Upward integration in the tire system

↑ Automobile "rolling" system
Tweel™
Tire plus wheel
Rubber polymers
Chemical monomers

If you're a tire producer and see this product, what goes through your mind? No more conventional tires! But maybe opportunities to supply this very strange looking tire "structure." Was Michelin talking to you about their plans? The rubber producer who's been "optimizing" their formulation to improve traction and reduce running resistance all of sudden finds that their phone stops ringing. If you were to supply this new "structure," how would your business change? Your manufacturing process? The types of people you hire? The raw materials you buy? The suppliers you talk to? The types of industry magazine you read? The types of industry meetings you attend?

What the patterns of inventions over decades tell us is that systems are inevitably integrated into their super-systems. So what *every* material, product, or system producer or supplier needs to be constantly thinking about is how the *function* that their product or system provides to their customer can be done *without* their product. Talk to and understand how their customer uses their product and *how that same function could be provided without their material, product, or system.* Then the question is: how does that change the fundamental nature of a business? Will it require an acquisition? A partnership? Licensing of new technology? Development of technology outside the norm?

Let's consider some additional examples of this concept of "upward system integration." What has happened to such items as "credit cards" and "cell phones"? They are beginning to merge into one functional device in the same way as cell phones and computers have done in the form of many different hand held devices. As discussed before, what has happened in the lawn care and pest control service industries? In many cases they are being done by the same company. This provides both additional conveniences to the homeowner and also flexibility in labor utilization by the companies.

Consider also the spread of huge discount stores run by Wal-Mart and Target where it is possible to buy virtually anything needed for a home including groceries, home maintenance, take-out food, gifts, and prescriptions. There will always be a market segment for people who enjoy small stores and less hustle and bustle, but the overall macroeconomic trend is consistent with this TRIZ line of evolution.

Exercises

1. Who is your customer? Why do they buy the product or service you sell? What function does it provide? If your customer was suddenly prevented from buying your product or service, what would they do (assuming there was no other supplier of your product or service)? How would their product or service change? How would your business change?
2. Make a list of all the products and services you buy from others. Now make the assumption that you cannot buy these products or services any more. What will you do?
3. Take the list from #2 and ask yourself how you could incorporate the function that this purchased service or product into your own in such a way that you did not need to buy the product or service anymore. What does your new product or service look like? What alternative materials might you consider? What advantages does it bring to the market place? What would its pricing structure look like? What new markets might be available? If we think about the Stanley Black & Decker Pivoting RapidRoller™ from this perspective, it might encourage amateurs to paint walls in a carpeted room since paint spills on carpets are much harder to clean than those on bare floors.
4. Think about your customer's customer. What would happen if they stopped buying the product or service provided by your customer?

Combining Upward Integration with Lines of Evolution

Let's now construct a simple "tic-tac-toe" diagram to take the next step in integrating this trend toward upward integration of products and systems with lines of evolution (Table 13.3).

Left to right would represent the lines of evolution, while moving upward would represent the upward integration of components into their super-system. The upper right hand corner would represent the most ideal system within these confines. Now of course this table could be different dimensions, especially in the upward direction since it is very easy to imagine many more than three systems and sub-systems. In the case of the "painting" example we have been using, the system could be a wall

Table 13.3 Strategy tic-tac-toe diagram

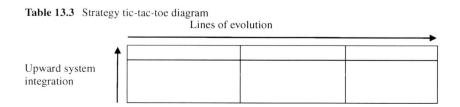

surface integrated into a room which is integrated into a house, which is integrated into a subdivision which in turn is integrated into a town or city. A tire is part of a wheel system which is in turn part of a steering/suspension system, which is part of a car, which is part of a transportation system, etc. We are just using a 3×3 matrix to get a basic understanding. Those of you with more complex systems are encouraged to go beyond the thinking in this section.

Let's consider the medical system we encounter on a regular basis. Looking at upward integration, what do we see? We have moved away from the single practitioner model to one where there are several physicians in an office as well as nurse practitioners. Lately we have begun to see doctors a part of hospital systems that in turn may be part of large health care organizations that may also be part of health care insurance organizations. If we look left to right across this diagram, what trends and lines of evolution do we see? Consider the field line of evolution. In assessing the state of a patient's health, we are now seeing the development of optical-and acoustical-based technologies which may replace blood tests and mechanically based "touchy feely" tests which relied on a physician's ability to sense what might be wrong.

Table 13.4 shows one possible 9-Box view of testing and evaluation. Again, upward system integration is vertical and progression along other lines of evolution is left to right.

If we were to look again at the "learning" industry, we see schools being integrated into the supersystem of all schools and learning resources which are now accessible through the use of advanced learning technologies that illustrate many of the lines of evolution such as web based devices, integrated devices, and devices which hide their complexity behind a user interface that appears simple to the user (Table 13.5).

You don't have to agree with what I have put in the boxes—think about this issue and put in your own input. Remember to think about all the lines of evolutions as well as the upward system integration while doing so.

Let's take a look at the general problem of "joining things." From a line of evolution standpoint, we have moved from mechanical clamping to thermally enhanced bonding to chemically based adhesives to magnetic clamping. If we look at upward integration, we see products that are pre-assembled and joined in their manufacture to minimize the need for additional joining. We see materials encapsulated in foams

Table 13.4 9-Box for medical testing

Second opinion from one other physician	Hospital system or insurance system review	Multiple opinions via web based sharing of information—automated expert system review
Single physician evaluates	Physician group reviews	Multiple physician group review
Mechanical test	Chemical test	Wave based test

Table 13.5 9-Box for education

Collection of schools in a school district paid for with property taxes	National network of customized learning options controlled by primary instructor	National network of customized learning with live peer interaction and automated change in course content and style based on progress of student
Conventional school	School plus customized home or remote learning	Computer based self-learning with live peer interaction
Teacher with 30 students	Teacher with aide and customized learning based upon knowledge of student needs	Computer based self-learning that adapts to needs of student

to make their assembly easier. This is frequently enabled by low cost labor in third world countries and it will be interesting to see what happens to this sort of assembly process as wages of those currently involved in this activity rise.

Exercises

1. Draw an empty 9-Box diagram and place the name of your product or service in the middle box. Now fill in the box above and below with the name of the materials you buy and, in the box above, the name of the product that your product is used within. (If it takes more than three levels of boxes to do this to complete the picture, go ahead and do it). Now ask yourself these questions:

   (a) How could you eliminate the need to buy the product you purchase from one or more suppliers (trimming, integration into the supersystem)?
   (b) How could your customer make their product without buying what they currently buy from you (you are trimmed)?

2. In that same box, look to the left and right in the context of the lines of evolution we have discussed. How did you accomplish the function that you currently do (how you produce your product or service) 10 years ago? How do you think you will make or produce your product or service 10 years from now (considering the lines of evolution such as dynamism and field evolution)? Was the movement from left to right to where you currently are a conscious one or were you dragged kicking and screaming? (Think of Kodak and electronic photography as an analogy). In thinking about the movement from where you are to where you might be 10 years from now, do you have the necessary skill set to get there? If not, how will you obtain them?

3. For your customer (and your customer's customer)—one and two levels up from you—do the same exercise. When they move from left to right, will your product or service be needed? If not, what are you going to do to maintain your position?

The TRIZ Cube

Let's take the thinking of the last two chapters a step further. Consider the start of Southwest Airlines. As the story goes, the start of this company was the drawing of a triangle using Dallas, Houston, and San Antonio on a napkin and the decision to price tickets not on the basis of undercutting what other airlines were charging at the time, but to look at the cost of automobile travel between these cities in terms of gas, oil, and other costs. This turned out to be significantly less than a typical 20 % discount from others' fares but opened up a huge market previously not recognized. The 9-Box diagram we have been discussing reflects only one view or system that can accomplish the result we want. If we want to get from Houston to Dallas, there is more than one way to do this and the 9-Box representing the airline approach is only one. Hitchhiking, walking, buses, and cars are others (in addition to the automobile). The original fares were priced to make it less expensive to fly than to drive a car. If we are not paying attention to other means of achieving the _function_ we are trying to accomplish, we may be surprised by a competitor in an area totally unknown to us. For example, if we go back to the heart stent and garbage bag example, for any 9-Box we draw, there are parallel ones representing other ways of achieving the same result. The lines of evolution we discussed apply to each parallel face of the cube as well.

There is more than one way of "decorating" or "covering" a wall. Wallpaper is one. What would happen to the paint business if we figured out how to make wallpapering user friendly and inexpensive? Pre-printed wall board is one example. Postal services have seen dramatic declines in volume as we use Email, video conferencing, and other means to accomplish the _function_, which is to communicate something to someone else. Why isn't the postal service in the Email business? Video conferencing business? If they saw their business as _communication_ rather than _delivery of letters_, would they be in financial trouble?

Here's a picture of a strategic cube, from a TRIZ perspective (Fig. 13.2).

The front face of the cube would represent your current business. Divided into nine boxes would segregate it into levels (bottom to top) and lines of evolution

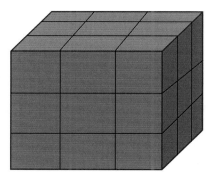

Fig. 13.2 The TRIZ Cube

(left to right). In a parallel 9-Box, we would show another approach to achieving the same _function_ and in an additional parallel box set would be an additional way to achieve that function. Each of these parallel universes has the same lines of evolutions and the same pull of upward integration. In our transportation example, we see the incorporation of video capabilities (TV's and DVD players) into cars that conceivably would allow video conferencing. Now we could debate the safety of this practice, but it's where we are headed. How would that affect airline travel? Current video conference technology?

There is a tremendous tendency to look only at our own products and processes (the center box in just one parallel face of the cube) and it's difficult enough to look to the left and right (lines of evolution) and up and down (system integration). It's even more difficult to think about parallel universes of accomplishing the same function. That's one of the reasons Southwest became one of the largest airlines in the USA. It's also why there are so many Internet based-document transfer services that are replacing overnight delivery of documents by FedEx® and UPS. The _function_ is to deliver documents not necessarily paper. In fact, it's the information on the document that's important, right?

Let's go back to painting walls again and think about this in the context of jargon and parallel approaches. Is the objective to paint the wall or _provide a decorative color or surface to the wall (or whatever the surface is)_? What parallel universe to painting walls is there? The wallboard manufacturers could conceivably come up with a way to precover their wallboard with the surface desired and eliminate household painting altogether. Now you are thinking to yourself all the reasons this isn't possible (that's a good idea, but…, remember?) and there are many, primarily having to do with protecting the surface from damage during house construction and wall installation. These are secondary problems that TRIZ could be applied to. What's a third parallel universe of getting a surface appearance (notice I did not say "paint") on the wall? Could it be done with lighting technology of some sort? Then the wallboard could still be white and no paint required. The "color" could be changed at will by the frequency of light wave chosen.

Exercises

1. Think about the product or service you provide. What are two other possible ways of providing the function that your product or service offers? (Not direct competitors!). What do you know about these other ways from a technical and business standpoint?
2. Assuming that you know the answer to #1, and your manner of delivering the product or service is being challenged, who would you acquire? Why?
3. Fedex® bought Kinko's®. Why?
4. ExxonMobil® just made a major acquisition of a coal company. Why?
5. What if you saw an announcement that General Motors just bought Intel®. Why?
6. What if you saw an announcement that Intel® just bought General Motors. Why?

7. Hertz® is looking for potential acquisitions. Who would they look at? What industries would they consider?

8. What is the next function that would be considered for incorporation into a cell phone?

9. You are a senior executive of a company which has just been awarded a contract to run a previously publicly run toll road. Who are your competitors? Why? Looking back at the "adding useful complexity" principle discussed earlier, what other services could you provide that would make this venture more profitable?

10. What is the next step along the "dynamism" line of evolution for you, your direct competitor, and at least one of your functional competitors? If you became proactive about this next step and none of your competitors were, what advantages would that provide? If you were not proactive, and they were, what competitive challenges would that present?

Part IV
Special Tools and Techniques, TRIZ Problem Modeling, and Integration of TRIZ with Other Tools

In the process of its development, TRIZ developed special tools, primarily focused on modeling problems. This was another step in the direction of being able to transfer known solutions from one area to another. Some of these have become the backbone of many TRIZ software products. In addition, as TRIZ became better known after its emigration to the West around 1990, it encountered existing creativity, innovation, and enterprise tools. In some respects TRIZ was viewed as a competitor; in others it was viewed in a complementary fashion. What we'll try to do in the next several chapters is to discuss briefly some of these modeling tools as well as discuss ways to integrate TRIZ with other enterprise tools.

Chapter 14
Special TRIZ Tools

There are a few unique TRIZ tools, not often used, but may resonate with you and be useful in special situations. They are "Smart Little People" modeling and using TRIZ "in reverse" when the problem we are trying to solve is one relating to a failure within a system. Let's take a look at each briefly.

(a) Smart Little People Modeling

Smart Little People Modeling (SLPM) in an approach to thinking about how an Ideal Result might be achieved. In this process, we use the analogy of smart little people (in some TRIZ literature you will see this technique described as "miniature dwarfs") to make a mental model and image of a problem. For example, let's imagine a problem in a desktop mouse which loses its capability to interact with a PC due to a buildup of dust and particulates from a dirty desk. If we imagined some miniature people inside the mouse or on the desk, what would they be doing? Well, they might be continuously scrubbing the surface contacting device. Or pushing the dust particles away. Or collecting and then disposing of the particulates (collecting and taking out the garbage?). Now with these mental images, how would we do this in the real world? If this method of envisioning this works for you, use it!

More detailed information on this process can be found in references [4, 11, 14].

(b) TRIZ in "Reverse"

Occasionally, the problem we are dealing with is a failure problem. This could be mechanical system which is unreliable and we do not understand the root cause (If we know the root cause and are not willing to fix it, this tool won't help, but if the proposed answer is too expensive or has some other defined drawback then the traditional TRIZ tools we have reviewed will help), a communication system

J. Hipple, *The Ideal Result: What It Is and How to Achieve It*,
DOI 10.1007/978-1-4614-3707-9_14, © Springer Science+Business Media New York 2012

that is not totally effective at all times (but these times are unpredictable and variable), or a product whose successful use by customers is not 100 % (and we don't understand why).

In this process, we invert part of the TRIZ algorithm. Instead of envisioning the Ideal Result, we invert and exaggerate the Ideal Result and, in effect, ask ourselves how we would make the product or system fail to perform or operate successfully 100 % of the time—the total opposite of the Ideal Result. We are asking the question "How?" instead of the normal question "What?" The mindset of people using this approach becomes proactive instead of reactive to a series of questions as is the case with tools such as FMEA or HAZOP. Once we identify the possibilities (which in my experience will always be a longer list than that generated by a checklist approach), we ask the TRIZ question, "where are the resources necessary to accomplish this negative result?" If we can't identify them, then this cause is removed from the list. If we can identify them, then we have identified a potential failure route and find a way to eliminate these resources.

The mental process and algorithm is as follows:

1. State the Ideal Result (the process or equipment works perfectly at all times)
2. Invert the Ideal Result (the process or equipment does not work perfectly at all times)
3. Exaggerate the inverted Ideal Result (the process or equipment _never_ works as intended). This is the key mental step as it puts people in an aggressive mental mind set. More possible causes for failure will be found in this mental state than reacting to a checklist. It's fun to be rewarded to deliberately make something go wrong!
4. Identify resources necessary to allow the causes identified in #3 to happen. Make sure that the resource lists discussed earlier are used to make sure all possible resources are identified. Eliminate any ideas that cannot be linked to available or potentially creatable resources.
5. If elimination of the resources cannot stop the failure from occurring, then we may have a problem that we solve with the 40 principles or separation principles. There may be a natural oscillation between steps 1, 2, and 5 in this process.

This TRIZ tool has been used in the analysis of mechanical equipment failures, software programming failures, electronic bank fraud, and chemical process releases.

In a chemical industry example of the use of this tool, a leak of a toxic gas from a chemical plant was caused the loss of scrubbing fluid that usually absorbed a toxic gas and prevented its release into the atmosphere. The typical approach to this, from a chemical industry HAZOP viewpoint, would be to look at all the control mechanisms, temperatures, pressures, flow rates, etc. and ask how we could control them better. This would typically involve the _addition_ of added control sophistication and equipment to prevent the loss of scrubbing fluid.

Using this "reverse" TRIZ approach, we would use the algorithm discussed previously as follows:

1. State the Ideal Result: The toxic gas is not released to the atmosphere.
2. Invert the Ideal Result: The toxic gas _is_ released to the atmosphere.
3. Exaggerate the inverted Ideal Result: The toxic gas is _always_ released to the atmosphere. Everyone in the town is killed and the plant manager is imprisoned. This is the key step—to exaggerate the inversion!
4. Identify resources necessary for this to happen. This is an interesting part of the problem. Many people when shown a flow sheet of the process will say things like "high pressure" or "high temperature," neither one of which will necessarily cause a leak. The resource that is necessary for a leak is a _hole_. Without a hole, there cannot be a leak, no matter what else is going on. In this particular process, the leak occurred at the top of a scrubber that had lost its scrubbing fluid. When the question was asked, "Why is the scrubber there?" the answer turned out to be to scrub nitrogen from a level monitoring system upstream. The question then became one of asking how else the level could be measured. If that could be done indirectly, then there was no gas to be scrubbed, no scrubber was needed, and the hole disappeared. Since the fluid whose level was being measured was paramagnetic, its level could be measured accurately by a system external to the tank, permitting the scrubber and its hole to be eliminated.

A HAZOP review, typically conducted on a process like this, does not always question the fundamental design and instead asks if the various design parameters such as level, temperature, pressure, and flow are not in their design range and then what are the consequences of these possible deviations. The reverse TRIZ approach basically changes the question to one of "how" instead of "what", putting participants in an aggressive mind set not requiring a large quantity of caffeine to keep them awake.

Chapter 15
TRIZ Problem Modeling

After Altshuller and his colleagues developed the contradiction table and inventive principles, work continued on ways to assist in the definition of problems in a general sense and to be able to graphically model problems as opposed to solely by contradictions. The first of these is what is commonly known TRIZ as Su-Field ("soo-field"), meaning substance field modeling. It is possible to graph a problem that has multiple contradictions and relationships in graphical form, the simplest of which would look like (Fig. 15.1).

What this diagram says is that the only way to have an interaction between the two substances (or objects or systems) is to have a field. Otherwise, the two substances will have no interaction. This field could be mechanical, thermal, chemical, etc. as we discussed in the lines of evolution section. One simple way of thinking about a system this simple is to ask these questions:

1. What other kind of field could be used? Would that have benefits? What is the cost? Can it be supplied by resources not previously considered?
2. Can the nature of the substances be enhanced or changed to enhance the effectiveness of the field currently in use? Think about the use of materials with different Curie points to enhance the effectiveness of a thermal and magnetic field.
3. Can an additional field be added to the system to increase the interaction between S_1 and S_2? This would change the diagram above to this (Fig. 15.2).

As with the concept of the 40 principles, there are a limited number of general problem models that exist (76 to be exact), and for each of them, there is a "standard" solution. This approach to problem modeling is the basis for many of the commercially available TRIZ software products. More details on this approach can be found in reference [11].

Another more general form of problem modeling is cause and effect modeling, again the basis for several TRIZ software products, but which can also be done by hand. It is also more useful for diagramming and modeling nontechnical problems in the business world.

J. Hipple, *The Ideal Result: What It Is and How to Achieve It*,
DOI 10.1007/978-1-4614-3707-9_15, © Springer Science+Business Media New York 2012

Fig. 15.1 Simple Su-Field
Model

Fig. 15.2 Su-Field Model
with 2 Fields

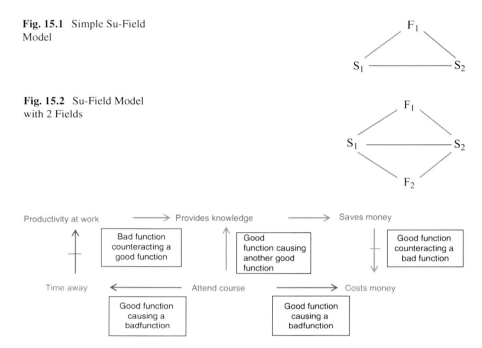

Fig. 15.3 A TRIZ cause and effect diagram

We can link two functions together with one of four links:

1. Causes or provides a good function ($\longrightarrow$)
2. Causes or provides a bad or negative function ($\longrightarrow$)
3. Counteracts a good function ($\longrightarrow\!\!\!+$)
4. Counteracts or prevents a bad function ($\longrightarrow\!\!\!+$)

An example of such a diagram, for someone traveling off site to take a TRIZ class, would look like this (Green functions are "good" and red functions are "bad") (Fig. 15.3).

When we look at a diagram of a problem done this way, the nature and colors of the arrows suggest some immediate TRIZ principles. A solid green line (good function causing another good function), we can ask:

1. How can we improve the good functions (make them *more ideal*)? What resources could we use?
2. What aspects of the good function could we trim and still obtain the result we want?
3. What is the next step in the appropriate lines of evolution that apply to this function? How could we use or implement?
4. We also see in this diagram a significant contradiction (a good function [attending the course]) causing both a good [providing knowledge] and bad function [Costing money, time away from work]. This is an opportunity to apply the contradiction table as well as the separation principles.

5. We see a bad function (time away from work) counteracting a good function (productivity). What principles could we apply to offset this?
6. We see a good function (saving money) counteracting a bad function (costing money). In what way could we improve this?

Note also that I have arbitrarily specified attending a TRIZ course as a "good" function. That's because I believe TRIZ is worth learning about. But someone who believes fervently in other creativity and innovation tools might feel otherwise, thus creating a very different diagram. This is no different, in principle, than the health care issues that we discussed earlier. It is a very enlightening exercise to segregate a problem-solving group (possibly based on function, position, or skill set) and ask each to diagram the problem as they see it, using either Su-field or cause and effect modeling. When the groups show each other their diagrams, it is frequently more valuable to discuss why the different groups envision the problem differently. How many contradictions did each identify? How many good and bad functions were identified? Why? What fields were identified as part of a system? Why do the different groups see this differently?

This type of diagramming can be done with software or with Post-it™ notes and is more useful for nontechnical or business problem solving.

Chapter 16
Using TRIZ with Other Tools

It is rare that an organization does not have some process for improving creativity and innovation, even if it's as simple as elementary brainstorming. On the sophisticated side, this could mean Six Sigma, Design for Six Sigma, QFD, and sophisticated consumer analyses. Assuming this, TRIZ entering an organization could be seen as competition or a complement to an existing process. Individuals who have been "certified" in these other tools and having invested a great deal of time and money in using and implementing these other tools and processes will justifiably ask, what is special about TRIZ? This sensitivity must be considered when trying to implement or evaluate TRIZ (or any other new innovation tool for that matter).

The basic answer to this question is that, in many cases, TRIZ provides the problem-solving tools that these other tools lack. If we look at QFD or Six Sigma, there is a lot of data taken to define the problem properly and then when it comes to solving the problem that has been identified or clarified, the suggestion is made to "brainstorm" for the answer. Here is where TRIZ can help. Whether it is the simple use of the 40 principles or identification of the contradictions that need to be resolved, TRIZ can make the problem-solving aspect of these other tools much more efficient and productive. TRIZ, for the most part, is not a competitor of these other approaches, but a complement to them.

Having said that, TRIZ has its role in problem definition as well. Tools such as QFD may do a superb job of understanding what current customers want and need, but do not necessarily assist in defining that parallel universe of companies or industries who can supply that same function in a different way. This is analogous to talking to the painting industry about their future requirements for paint pans while the elimination of paint pans is what is being planned and you are not being told. QFD has a step in its process that specifically identifies contradictions that stand in the way of achieving the quality that is needed. Obviously, resolving contradictions through the contradiction matrix and separation principles can be used here.

We discussed earlier the difference between a TRIZ and Six Sigma approach to a general problem. However, if we had a problem defined by Six Sigma, we could

J. Hipple, *The Ideal Result: What It Is and How to Achieve It*,
DOI 10.1007/978-1-4614-3707-9_16, © Springer Science+Business Media New York 2012

use TRIZ as a more focused problem-solving process as opposed to brainstorming. TRIZ would ask questions relating to the Ideal Result (Maybe 7 sigma? Looking at the supersystem and eliminating the problem altogether?), and a more thorough look for resources that are available to solve the problem. The focus of Six Sigma on the problem at hand may prevent us from looking at the systems above and below for resources. You have seen how TRIZ broadens the definition of "resources" and how it expands our view of what may be available to solve the problem at hand. If the resolution of a contradiction is what is required to get to a Six Sigma quality level, then TRIZ provides the tools to accomplish that. In Design for Six Sigma (DFSS), we are trying to design a process or product which will have Six Sigma performance before it goes into operation or use. In this area, all of the TRIZ tools can be used. Ideal Result thinking and resource identification and utilization can assure that the best concepts are being considered in the first place. Any early stage design contractions can be analyzed with the TRIZ contradiction table and the TRIZ separation principles. The lines of evolution can help design personnel think about the next generation product or process earlier than their competitors.

In some of the more psychologically based techniques such as Creative Problem-solving or DeBono's Lateral Thinking® and Six Thinking Hats® the TRIZ principles can be used to focus the brainstorming processes as well as provide more structure to the problem definition. A simple technique to supplement these processes is to ask, "Who else has a problem like this? How they solved it?" The DeBono technique of using the word "po" (meaning provocative operation) to inject an unusual idea can be greatly assisted with the simple technique of an exaggerated Ideal Result statement. Then look for resources to accomplish this "out of the box" idea. More information on the use of TRIZ with other processes is available [1].

One of the more difficult issues that you may run into, discussed previously, is trying to use TRIZ concepts when an existing creativity process already exists. In the case of a process whose basis is to generate large quantities of ideas, you might suggest that more time be spent on problem definition, possibly using one of the TRIZ modeling tools to be discussed in the next chapter. Another approach is to try to get the discussion focused around a contradiction that needs to be resolved which would allow you to introduce the contradiction table and the separation principles in a non-threatening way. Another approach, mentioned previously, might be to use the 40 principles as a random stimulus mechanism. The principles could be put on cards and then randomly distributed and each participant asked to suggest an idea based on the principle they have. When this exercise generates some interesting ideas, the question might come up, "Where did these principles come from?" and then a TRIZ discussion could start that might lead to the group wanting to learn more.

The worst thing that can be done is to suggest that the currently used tools have no value. *Any* tool or process that helps us to create new ideas or view a problem differently has value. If the TRIZ journey needs to start by attaching itself to an existing process, start there and move on!

Exercises

1. You have just finished a "house of quality" QFD exercise and have identified several contradictions between customer requirements and your capability to meet them. What TRIZ tools will use to deal with this problem?
2. Your Six Sigma project has reached a dead end in terms of its ability to get beyond a "4-sigma" deviation. What TRIZ tools come to mind in addressing this challenge?
3. How could you use "upward integration" as a thinking concept to help you deal with both (1) and (2)?
4. How would you explain to a certified practitioner of either DeBono's tools or some version of brainstorming what is different about TRIZ and the value it might have?

Reference

1. Hipple J (2005) The integration of TRIZ with other ideation tools and processes as well as with psychological assessment tools. Creativity Innov Manag 14(1):22–33

Summary

As we have seen, TRIZ is a methodical approach to innovation, creativity, and problem solving. It, in effect, brings all of the inventors in the world into the room with you. It requires that the problem owner give up their ego and be willing to admit that their problem, in a general sense, has already been solved. It requires significant up front problem definition time to match the problem with existing solutions. Not being able to give up one's ego and the unwillingness to spend time defining a problem adequately are the two biggest barriers to the use and implementation of TRIZ tools within an organization.

I have found it very useful, as indicated in how I have presented the material, to use the most basic of the TRIZ tools first. It has been amazing to me, over the years, to see how often a simple web search around a more general description of the problem can produce some novel ideas and approaches to organizations that truly think they have a unique problem that no one else has ever had. In addition to the air traffic control display example presented earlier, there is another case where a need to maintain exact "parallel to the ground" stability in a medical instrument was mapped against the same challenge in the motion picture industry. There are really no new problems—only new applications of known problem-solving principles to new problems.

The TRIZ lines of evolution provide the same insights into where a product, business, or service may be heading in the future. These lines contain many discontinuities. If these are recognized ahead of time, an organization can change its focus, the type of people it hires, and who it acquires or merges with so as not to be surprised. They allow an organization to get ahead of the curve in terms of its strategy and plans behind the strategy.

J. Hipple, *The Ideal Result: What It Is and How to Achieve It,*   181
DOI 10.1007/978-1-4614-3707-9, © Springer Science+Business Media New York 2012

The concept of contradictions is very powerful. Thinking about *resolving* contradictions, instead of compromising around them, is the key to breakthrough new products, businesses, and services. Running *toward* a contradiction is the fastest way to keep ahead of the competition which is trying to "optimize" around the problem. TRIZ provides many ways of resolving contradictions.

At its roots, TRIZ is an algorithm, flying in the face of the psychological approaches using randomness as the key to breakthrough inventions. An algorithm is a path to a solution. If the path is followed, you will arrive at the destination.

Epilogue

A vice president of a product design firm, after reading about TRIZ, commented:

> It seems like TRIZ is trying to create an equation for innovation… I think it's a great aspiration. But if there's an equation for innovation out there, your competitor can do the same—which means the competitive challenge can easily be lost.

Well this may be true, but if you're the only one who has seen the light, you have a great advantage. One of the biggest barriers to innovation is the status quo and peoples' egos. Many businesses have failed due to both of these. Don't let these get in your way of your innovation journey.

J. Hipple, *The Ideal Result: What It Is and How to Achieve It*,
DOI 10.1007/978-1-4614-3707-9, © Springer Science+Business Media New York 2012

Trademarks

Six Thinking Hats® and Lateral Thinking® are registered trademarks of APTT

Rain-X® is a registered trademark of ITW Global Brands

Black and Decker Pivoting RapidRoller™ is a registered trademark of Stanley Black & Decker

Listerine® is a registered trademark of McNeil-PPC, Inc.

Target® is a registered trademark of Target Stores

Tweel™ is a registered trademark of Michelin Corporation

WebEx™ is a registered trademark of Cisco Corporation

GoToMeeting® is a registered trademark of Citrix Corporation LLC

govino® is a registered trademark of By The Glass LLC

Nature's Bounty™ is a registered trademark of Nature's Bounty, Inc.

Space Bag® is a registered trademark of ITW Space Bag

Scrubbing Bubbles® is registered trademark of S.C. Johnson, Inc.

CVS™ is a registered trademark of CVS Corporation

Sears® Kenmore® is a registered trademark of Sears Brands LLC

WiSE™ and AirSage® are registered trademarks of AirSage, Inc.

Acuform™ and Depomed® are registered trademarks of Depomed, Inc.

UPS™ is a registered trademark of United Parcel Services of America, Inc.

FedEx™ is a registered trademark of FedEx, Inc.

Glide™ is a registered trademark of Proctor and Gamble, Inc.

DietPlate® is a registered trademark of Slimware, Inc.

Fresh&go™ is a registered trademark of Fresh & Go, Inc.

Excedrin® is a registered trademark of Novartis Human Health, Inc.

Perpetual kid™ is a registered trademark of perpetual kid.com

Dyson® is a registered trademark of Dyson, Inc.

X-Beam® is a registered trademark of Apex Tool Group

Xiameter® is a registered trademark of Dow Corning Corp.

Terminix® is a registered trademark of The Terminix Co. Limited Partnership

HBDI® is a registered trademark of Inscape, Inc.

Fedex® is a registered trademark of Fedex Corp.

Post-it Notes® is a registered trademark of 3M Corporation

J. Hipple, *The Ideal Result: What It Is and How to Achieve It*,   185
DOI 10.1007/978-1-4614-3707-9, © Springer Science+Business Media New York 2012

Resources and Additional Reading

1. Altshuller G (1996) And suddenly the inventor appeared. Technical Innovation Center
2. Altshuller G (1999) The innovation algorithm. Technical Innovation Center
3. Altshuller G (1988) Creativity as an exact science. Gordon and Breach
4. Belski I (2007) Improve your thinking. TRIZ4U
5. Domb E, Rantanen K (2002) Simplified creativity. St. Lucie Press
6. Fey V (2005) Innovation on demand: new product development using TRIZ. Cambridge Press
7. Fey V, Clausing D (2004) Effective innovation: the development of winning technologies. ASME Press
8. Hipple J (2005) The integration of TRIZ with other ideation tools and processes as well as with psychological assessment tools. Creativity Innov Manag 14(1):22
9. Hipple J (2008) Predictive failure analysis: planning for the worst so that it never happens. Fam Commun Health 31(8):71
10. Hipple J (2005) Solve problems inventively. Chemical Engineering Progress, April, May, June
11. Mann D (2003) Matrix 2003. Creax Press
12. Mann D (2004) Comparing the classical and new contradiction matrix. TRIZ Journal
13. Mann D (2010) Matrix 2010. IFR Press
14. Mann D (2010) Hands on systematic innovation. IFR Press
15. Salamatov Y (1999) TRIZ: The right solution at the right time. Insytec
16. Savransky S (2000) The engineering of creativity. CRC Press
17. Innovation-TRIZ: http://www.innovation-triz.com
18. ASME/AIChE: http://www.asme.org/products/courses/triz-the-theory-of-inventive-problem-solving
19. Altshuller TRIZ Institute: http://www.aitriz.org
20. Wikipedia: http://en.wikipedia.org/wiki/Structure_and_function_of_TRIZ

Templates for Your Use

1. Optimization Graph

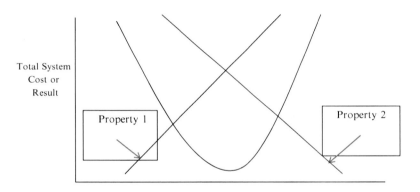

Total System
Cost or
Result

Property 1

Property 2

System Optimization Template

2. Different Views of the Ideal Result

Stakeholder #1 _____

Stakeholder #2_____

Stakeholder #3_____

Stakeholder #4_____

What percentage of each stakeholder's Ideal Result has been satisfied?

Stakeholder #1 _____

Stakeholder #2_____

Stakeholder #3_____

Stakeholder #4_____

J. Hipple, *The Ideal Result: What It Is and How to Achieve It*,
DOI 10.1007/978-1-4614-3707-9, © Springer Science+Business Media New York 2012

3. Substances and Material Resources

What		How might you use
1.	_____	_____
2.	_____	_____
3.	_____	_____
4.	_____	_____
5.	_____	_____
6.	_____	_____
7.	_____	_____
8.	_____	_____
9.	_____	_____
10.	_____	_____

4. Space Resources

Type of space	How can you use?	What affects? In what way?
1. Voids	_____	_____
	_____	_____
2. Surfaces	_____	_____
	_____	_____
3. Lengths	_____	_____
	_____	_____
	_____	_____

5. Time Resources?

Type of time	How to use?
1. Time prior to your supplier receiving the material used to make your raw materials	_____
2. Time during your suppliers' processes	_____
3. Inventory time at your suppliers	_____
4. Transit time from your suppliers	_____
5. Time in your receiving area	_____
6. Time not used completely within your process	_____
7. Time after your process	_____
8. Time in your warehouse	_____
9. Transit time to your customers	_____
10. Time in your customers' process	_____

6. Informational Resources

Functions being performed	Information being generated
_____	_____
_____	_____
_____	_____
_____	_____
_____	_____
_____	_____

7. Fields and Field Conversions

Field	Field conversions	Useful for?
1. _____	_____	_____
2. _____	_____	_____
3. _____	_____	_____
4. _____	_____	_____
5. _____	_____	_____
6. _____	_____	_____
7. _____	_____	_____
8. _____	_____	_____

8. Negative Resources

"Negative" aspects		Use this for
1.	_____	_____
2.	_____	_____
3.	_____	_____
4.	_____	_____

9. Trimming Table

Part of a system to be eliminated	What does it do?	What else in the system could do this function?
_____	_____	_____
_____	_____	_____
_____	_____	_____
_____	_____	_____
_____	_____	_____

10. 40 Inventive Principles

1. Segmentation
2. Taking Out/Separation
3. Local Quality
4. Asymmetry
5. Merging
6. Universality
7. Nested Doll
8. Anti-weight
9. Preliminary Anti-action
10. Preliminary Action
11. Beforehand Cushioning
12. Equipotentiality
13. "Other Way Around"/Do It in Reverse
14. Spheroidality/Curvature
15. Dynamization
16. Partial or Excessive Action
17. Another Dimension
18. Mechanical Vibration
19. Periodic Action
20. Continuity of Useful Action
21. Skipping
22. "Blessing Disguise"
23. Feedback
24. Intermediary
25. Self-service
26. Copying
27. Cheap Short Living Objects
28. Mechanics Substitution
29. Pneumatics and Hydraulics
30. Flexible Shells and Thin Films
31. Porous Materials
32. Color Changes
33. Homogeneity
34. Discarding and Recovering
35. Parameter Changes
36. Phase Transitions
37. Thermal Expansion
38. Strong Oxidants
39. Inert Atmosphere
40. Composite Materials

11. Traditional Contradiction Table

The traditional contradiction table can be accessed on the web at:

https://www.innovation-triz.com
https://www.aitriz.org
https://www.triz40.com

12. Group Answers to the Contradictions in Innovation Problem

Innovation contradiction	Resolution idea suggested by separation in
Someone's job versus everyone's job	Time: Rotate responsibilities, allocate % time, some full-time personnel
	Space: Innovation space, room, or lab "Kindergarten" room
	Parts and whole: Targeted focus groups, innovate within parts of a project, break company into separate businesses, focus on parts of the project which really require innovation
	Upon condition: Simulate customer and business emergencies, groups for idea generation vs. evaluation
Inside business versus outside business focus	Time: Use life cycle analysis, use "outside in" as the analysis focus vs. "inside out"
	Space: "Mix it up"—have different locations for groups focusing in these two different areas
	Parts and whole: Consider internal ventures or separate business structure to deal with totally new businesses
	Upon condition: Change nature of innovation with business cycle, outside innovation always separate
Passion versus objectivity	Time: Increase objectivity as financial commitment rises, use outside reviews as needed
	Space: Objectivity from outside, passion from within
	Parts and whole: Separate types of reviews, perspectives from different divisions
	Upon condition: Separate idea generation from evaluation
Chaos versus discipline	Time: Chaos to discipline as project progresses, Separate project reviews from "brainstorming" sessions
	Space: Safety zones for chaos
	Parts and whole: External participants, maintain small group to continuously challenge, parallel group to continue to innovate primary project is progressing
	Upon condition: Change as a function of the project, customer, business environment, suppliers
Risk versus job security	Time: Identify gates and break points well, use senior people (with no fear of job loss!) to assess
	Space: Secure space for frank comments, keep NPD team isolated from normal business activities
	Parts and whole: Competitive partnering, obtaining 360 degree feedback
	Upon condition: Gate review processes as a function of monetary needs, defined incremental commitments make sure NPD teams are rewarded expeditiously

13. Lines of Evolution Templates

How "ideal" is your product, service, or system? Remember the definition: Something performs its function and doesn't exist or mathematically, the ratio of the good functions provided divided by the bad functions or negative side effect.

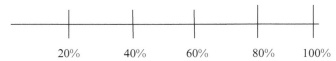

20%   40%   60%   80%   100%

How well are you using the resources in your system? Remember to look above and below your product, business, or service in the food chain.

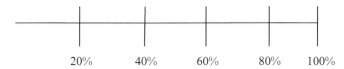

20%   40%   60%   80%   100%

How dynamic and responsive is your product, service, or system?

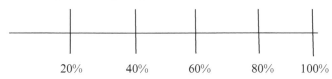

20%   40%   60%   80%   100%

Is your product, system, or service too simple or too complex?

How do you know? How did you decide? If it is too simple, what useful complexity can you add? If it's too complex, how can you trim and simplify without losing functionality? How can you trim and simplify and reduce functionality in a way that will attract customers?

In what way is your product, service, or system too complex?

_____
_____
_____
_____

What could you "trim" from your product, service, or system?

_____
_____
_____
_____

What would the impact be?

_____
_____
_____
_____

What are the primary fields used in your product, system, or service?

_____
_____
_____

Given the known progression of field evolution as mechanical/thermal/acoustic/chemical/electronic/electromagnetic, what is the next field along this line of progression for your product, system, or service?

_____
_____
_____
_____

What do you know about it and how to apply it?

_____
_____
_____
_____

Index